**A TEXT BOOK OF**

# NANO ELECTRONICS

**(Program Elective Course – 2)**
**FOR**
**SEMESTER – VI**

## THIRD YEAR (T.Y.) B. TECH COURSE IN ELECTRONICS / ELECTRONICS & TELECOMMUNICATION ENGINEERING

**Strictly According to New Revised Credit System Syllabus
of Babasaheb Ambedkar Technological University (BATU),
Lonere, (Dist. Raigad) Maharashtra,**
(w.e.f. June 2019-20)

**Dr. S. T. Patil**
M.Tech. (CSE), Ph.D. (Computer)
Professor,
Dept. of Computer Engineering,
Vishwakarma Institue of Technology
Bibwewadi, PUNE.

N1121

**NANO ELECTRONICS (E&TC, BATU)**                    **ISBN : 978-93-89825-10-7**

| | | |
|---|---|---|
| **First Edition** | : | **December 2019** |
| © | : | **Author** |




**Published By :**
**NIRALI PRAKASHAN**
Abhyudaya Pragati, 1312 Shivaji Nagar
Off J.M. Road, PUNE 411005
Tel : (020) 25512336/37/39
Email : niralipune@pragationline.com

## ➢ DISTRIBUTION CENTRES

**PUNE**

**Nirali Prakashan (Local)** : 119 Budhwar Peth, Jogeshwari Mandir Lane, Pune 411002, Maharashtra
Tel : (020) 2445 2044, Mobile : 9657703145, Email : niralilocal@pragationline.com

**Nirali Prakashan (Outstation)** : S. No. 28/27 Dhayari, Near Asian College, Dhayari, Pune 411041, Maharashtra
Tel : (020) 2469 0204, Fax : (020) 2469 0316, Mobile : 9657703143
Email : bookorder@pragationline.com

**MUMBAI**

**Nirali Prakashan** : 385 S.V.P. Road, Rasadhara Co-op. Hsg. Society Ltd., Girgaum, Mumbai 400004, Maharashtra
Tel : (022) 2385 6339 / 2386 9976, Fax : (022) 2386 9976, Mobile : 9320129587
Email : niralimumbai@pragationline.com

## ➢ DISTRIBUTION BRANCHES

**JALGAON**

**Nirali Prakashan** : 34 V. V. Golani Market, Navi Peth, Jalgaon 425001, Maharashtra
Tel : (0257) 222 0395, Mob : 94234 91860, Email : niralijalgaon@pragationline.com

**KOLHAPUR**

**Nirali Prakashan** : New Mahadvar Road, Kedar Plaza 1$^{st}$ Floor, Opp. IDBI Bank
Kolhapur 416012, Maharashtra. Mobile : 9850046155
Email : niralikolhapur@pragationline.com

**NAGPUR**

**Nirali Prakashan** : Above Maratha Mandir, Shop No 3, Second Floor,
Rani Jhanshi Square, Sitabuldi, Nagpur 440012, Maharashtra
Tel : (0712) 254 7129, Email : niralinagpur@pragationline.com

**DELHI**

**Nirali Prakashan** : 4593/15 Basement, Agarwal Lane, Ansari Road, Daryaganj
Near Times of India Building, New DelhiV 110002 Mobile : 8505972553
Email : niralidelhi@pragationline.com

**BENGALURU**

**Nirali Prakashan** : Maitri Ground Floor, Jaya Apartments, No. 99, 6$^{th}$ Cross, 6$^{th}$ Main,
Malleswaram, Bengaluru 560003, Karnataka
Mobile : 9449043034, Email : niralibangalore@pragationline.com



niralipune@pragationline.com  |  www.pragationline.com
Also find us on  www.facebook.com/niralibooks

# PREFACE

It gives me great pleasure to present the book **"Nano Electronics" (Program Elective Course – 2)** for the students of **Semester VI Third Year (T.Y.) B. Tech. Course Electronics / Electronics and Telecommunication of Dr. Babasaheb Ambedkar Technological University (BATU), Lonere, Dist. Raigad (Maharashtra).** This book is strictly as per the new revised syllabus 2019-20 Pattern, effective from the Academic Year July 2019-20.

In New Revised Syllabus, there will Class Assessment (CA) 20 Marks, Mid Sem. Exam. (MSE) 20 Marks and End Sem. Exam. (ESE) 60 Marks. End Sem. Exam. will be based on all Six units and each unit will carry 20 Marks.

**The Theory Course will have 3 Credits.**

The basic objective of this book is to bridge the gap between the vast contents of the reference books, written by the renowned International Authors and the concise requirements of Undergraduate Students. This book has been written in a comprehensive manner using Simple and Lucid language, keeping in mind students' requirements. The main emphasis has been given on exploring the basic concepts rather than merely the Information. Solved Examples and Exercises have been provided throughout the book and at the end of the Unit. Also I have given **Model Question Papers** for practice at the end of book.

My special thanks to my family members, students and all those who directly or indirectly supported me in this project.

I also take this opportunity to express my sincere thanks to Shri. Dineshbhai Furia, Shri. Jignesh Furia, Mrs. Nirali Verma, Shri. M. P. Munde and entire team of Nirali Prakashan, namely Mrs. Deepali Lachake (Co-ordinator), and her colleagues who really have taken keen interest and untiring efforts in publishing this text.

The advice and suggestions of my esteemed readers to improve the text are most welcome and will be highly appreciated.

**Pune**                                                                                         **Author**

# SYLLABUS

## Unit 1 : Overview Nano Technology

Introduction to nanotechnology, Nano devices, Nano materials, Nano characterization, Definition of Technology node, Basic CMOS Process flow, meso structures.

## Unit 2 : Basics of Quantum Mechanics

Schrodinger equation, Density of States, Particle in a box Concepts, Degeneracy, Band Theory of Solids, Kronig-Penny Model. Brillouin Zones

## Unit 3 : MOS Scaling Theory

Shrink-down approaches: Introduction, CMOS Scaling, The nanoscale MOSFET, Finfets, Vertical MOSFETs, limits to scaling, system integration limits (interconnect issues etc.)

## Unit 4 : Nano Electronics Semiconductor Devices

Resonant Tunneling Diode, Coulomb dots, Quantum blockade, Single electron transistors, Carbon nanotube electronics, Band structure and transport, devices, applications, 2D semiconductors and electronic devices, Graphene, atomistic simulation

## Unit 5 : Properties of Nano Devices

Vertical transistors, Fin FET and Surround gate FET. Metal source/drain junctions – Properties of schottky functions on Silicon, Germanium and compound semiconductors - Work function pinning.

## Unit 6 : Characterization Techniques for Nano Materials

FTIR, XRD, AFM, SEM, TEM, EDAX Applications and interpretation of results, Emerging nano material, nano tubes, Nano rods and other Nano structures, LB technique, Soft lithography Microwave assisted synthesis, Self-assembly.

# CONTENTS

# OVERVIEW NANO TECHNOLOGY

## 1.1 INTRODUCTION TO NANOTECHNOLOGY

- The word Nanoelectronics consists of two parts:
  1. Nano and
  2. Electronics.

- One nanometer (nm) equals 0.000 000 001 metre. This very tiny length goes almost beyond our imagination and cannot be seen using traditional microscopes. A nanometer to a meter is the same as a marble compared to the size of the earth.

- A hair is around 100.000 nm thick. Nanoelectronics studies the flow of charge through various devices and materials such as semiconductors, metals, resistors, inductors and nano–structures. Electronic parts are important for many tools we use in daily life: cell phones, computers, television, home equipment sand cars. Nanoelectronics is the branch of electronics dealing with miniaturised electronic circuits integrated on semiconductor 'chips'.

- Its basic element is the transistor. Up to recently, transistor dimensions were in the micrometer range but today they are manufactured at 65 nanometer or below. Nanoelectronics research is dealing with devices of dimensions of 22 nm and smaller, the technology and equipment to make them, the competences to design them with integrated circuits and electronic components and the art to manufacture them cheap in large volumes. The smaller the technology, the smaller, faster, more powerful and cheaper electronic equipment can be.

- The scientists have established a universal system for the classifying the size, it is typically based on the metric system with the base unit as meter. The metric system has a prefixes size of the centimeters. 100 or 0.01. It is expressed as $10^{-2}$. The Nano is third smallest increment described on the metric scale its value 0.00000001 or 1 billion base value. Now–a–days nano–technology and nanoelectronics used to interchangeably. They have a number of innovations of the modern ability to work and see with very small units. This small units are on atomic or molecular scale.

- Nananoelectronics signifies to the products of the evolutionary development of silicon–based micro electronic transistor technology categorized by the miniaturization and increased integration of elements which does not indicate instrumental implementation of quantum size effects.

- Nananoelectronics is also applied to a combination of electronic instruments, devices and their production techniques based on new effects.

- The scale of several tens of nanometers, the specific sizes of elements become comparable to certain fundamental physical values, such as, shielding distance, electron path, de Broglie wavelength, which involves the emergence of new physical effects and the existence of some fundamental physical restrictions of the capabilities of such devices. This is how nanoelectronics differ from microelectronics, which depends on the macroscopic laws of classical physics.

- The modern electronics are on the turn–around point in materials. The past, Al or $SiO_2$ which has an attraction to Si has been used and the functionalities are conveyed by manufactured devices.

- Consequently, the device structures have been intricate progressively and now it almost closes to the fabrication limit. Therefore, for the future Nano device needs new materials which give superior performance in Nano scale.

- The future Nano devices comprising LSI consist of a lots of interfaces with numerous materials. Basically nanoelectronics materials invented by the silicon, gallium, arsenic, and others, all these materials, Silicon is one of the most commonly available element on the earth. Silicon is cheaper, abundant and it requires low power only. Nanotechnology gives the more consistent small size devices and it gives faster working.

- The emerging potential for the science of nanoelectronics can be used in development of medicine technology.

**Some of the Several Perspectives to the Concept of Nanoelectronics are:**

1. Nanoscale dimensions of nanoelectronic components allow for systems of giga–scale complexity measured in terms of component on a chip or in a package. This scaling feature and the giga–scale systems can be described as the 'More Moore' domain of development.

2. Another is that nanotechnology is very diverse and allows the integration of purely electronic devices with mechanical devices, bio–devices, chemical devices, etc. Also, digital systems can be combined with analog/RF circuits. This technology fusion can be described as the 'More than Moore' domain of development.

3. A third is that traditional scaling limits in standard CMOS technology are reached during the next decade, fundamentally, new nanoscale electronic devices. This development of nanoelectronic components can be denoted the 'Beyond CMOS' domain of development.

- Over the years, semiconductor technologists have pushed transistors to smaller and smaller feature sizes. This steady shrinking of device size has resulted in an information revolution that today impacts on virtually every facet of our lives. Even though transistors are close to reaching their ultimate size limitation relative to integrated circuit performance, the stunning achievements in fabrication tool development in the last several decades as part of the microelectronics revolution now allow for molecular–level structural tailoring of materials not available or explorable except through naturally occurring atomic processes. Indeed, the age of nanoscience has arrived.

- Nanoscience is the study of material manipulation and its intended physical consequences at the molecular scale, that is, on a scale of the order of a few hundred Angstoms less than one thousandth of a human hair. The extraordinary feature of nanoscience is that it allows for the tailoring and combining of the physical, biological, and engineering properties of matter at a common level of manipulation and control; this feature provides an enormous opportunity to fabricate, tailor, and embed novel, specifically targeted chemical, biological, physical, and material attributes at the lowest level of material building block. The challenge of nanotechnology is to scale up or suitably package nano–based material concepts to a robust, usable macroscopic level while preserving the desired embedded nano features.

- The early techniques of molecular and cluster beam epitaxy, and more recently, the chemistry of self–assembly and molecular design, have facilitated the fabrication of structures with atomic layer resolution; as well, advanced lithographic and replication methods have provided the capability for defining lateral dimensions with an accuracy of Angstroms.

- Over the evolving years of progressive microelectronics, these revolutionary nanofabrication techniques have ushered microelectronics into the nanoelectronic regime by providing quantum wells, wires, and dots to serve as the basic workhouse structures for the study of many novel quantum phenomena and device concepts.

- Today, with new and emerging chemical processes, nanoscience has the potential to provide a unique spectrum of new concepts and capabilities for future generations of electronics as well as other areas including nanoelectromechanical structures, designer functional materials and textile fabrics, medical sensors and probes, and the like.

- This potential will be realized through a systems–level approach which concurrently integrates nano–embedded properties with new paradigms for device physics, flexible architectures, and hierarchical design principles.

- The perspective includes a discussion of the need to transition from the classical to the quantum picture of nature, a view of life in the nano lane, a discussion of quantum engineering – the application of quantum principles to nano objects including some of the author's own interests, an observation concerning a materials explosion toward new applications of nanoscience and technology, and a discussion of future directions relevant to nanoscience and engineering.

- Nanoelectronics refers to the use of nanotechnology in electronic components. The term covers a diverse set of devices and materials, with the common characteristic that they are so small that inter-atomic interactions and quantum mechanical properties need to be studied extensively. Some of these candidates include: hybrid molecular / semiconductor electronics, one-dimensional nanotubes / nanowires (e.g. Silicon nanowires or Carbon nanotubes) or advanced molecular electronics.

- Nanoelectronic devices have critical dimensions with a size range between 1 nm and 100 nm. Recent silicon MOSFET (metal-oxide-semiconductor field-effect transistor, or MOS transistor) technology generations are already within this regime, including 22 nanometer CMOS (complementary MOS) nodes and succeeding 14 nm, 10 nm and 7 nm FinFET (fin field-effect transistor) generations. Nanoelectronics are sometimes considered as disruptive technology because present candidates are significantly different from traditional transistors.

## 1.2 BAND STRUCTURES IN SILICON

- Silicon is a type of semiconductor material in which number of free electrons is less than conductor but more than that of insulator. Silicon has an enormous application in the field of electronics, for having this unique characteristic.

- There are two types of energy bands in silicon which are conduction band and valance band. A series of energy levels having valance electrons forms the valance band in the solid.

- At absolute temperature (0°K) the energy levels of the valance band is filled with electrons. When the electrons are in valance band, this band consists of extreme amount of energy, no current flows due to such electrons.

- As shown in Fig. 1.1 energy band of silicon levels of energies of electrons in the material. there are two types of energy bands one is conduction band and second is valance band. Valance electrons band with highest energy level. Free electrons are in conduction band with minimum amount of energy.

- Valance and conduction bands are separated by the amount of energy known as the forbidden energy gap. This amount is nearly 1.2 eV at 300° K. In intrinsic silicon, the Fermi level lies in the middle of the donor atoms, it becomes n–type when Fermi level moves higher, i.e. closer to conduction band.

- When intrinsic silicon is doped with acceptor atoms, it becomes p – type and Fermi level moves in the direction of valance band.

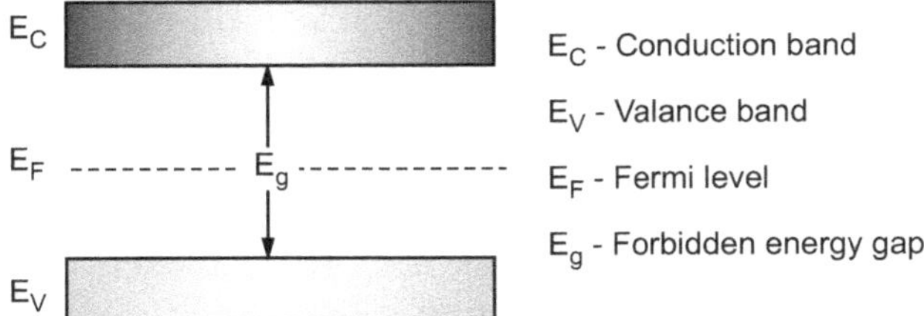


**Fig. 1.1: Energy bands in Silicon**

- Fabrication of Silicon nanoelectronics materials plays two important things, that is, quantum effect and single electron effect.

- The advantages of fabrication of silicon nanoelectronics are to decrease the size and power consumption.

- Fabrication of silicon nanoelectronics involves the quantum dot, quantum wire, quantum wel is and single electron transistor device (nano MOS) used to the lower power consumtion.

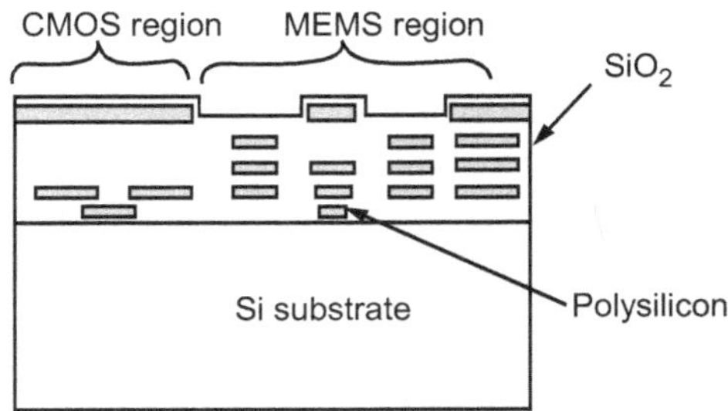


**Fig. 1.2: Silicon based nano CMOS**

- Especially for silicon, semiconductor band structures are hard to describe with an analytical formula. The plot is drawn for energy values along particular edges.

- The energy dispersion along the straight line from point T to point X, which is called $\Delta$ line. For silicon the conduction band minima lie on the six equivalent $\Delta$ lines along (100) directions and occur at about 0.85% of the way to the zone boundary. These are the well-known, equivalent ellipsoidal constant energy valleys. When electrons gain 0.13 eV of energy, they can cross the zone boundary.

- There is a second minimum in the first conduction band at point L , which lies 1.0 eV above the first valley. The second conduction band valley is only 0.1 eV above the minimum of the first conduction band.

- Carriers above 0.1 eV in kinetic energy may reside in either of the two conduction bands before the second valley of the first band is occupied. Under large electric fields, electrons populate the entire Brillion zone and the band structure at energy minima cannot be described by simple analytical approximations.

- The valence band maximum for the heavy–hole, light–hole and for the split–off band is positioned precisely at the T–point. It is clear that for a small electric field the concentration of holes is higher in the region around the T–point. It is thus important to use, for full band Monte Carlo simulations, a discretization of the momentum space, which demonstrate a higher mesh density around the T–point.

- The complexity of semiconductor band structures forces different approximations to reduce intensity and computational costs. Two analytical models, namely the parabolic and the non–parabolic model are widely used.

- The FBMC method produces a more general description, suitable also for higher energy values. This description is commonly based on the pseudo potential method.

## Analytical Approximations

Silicon band structure simplest model is centered on the effective mass. If the band structure is identified, the energy wave vector relation E(k) in one dimension can be stretched in a Taylor series as:

$$E(k) = E(0) + \left.\frac{\partial E(k)}{\partial k}\right|_{k=0} k + \frac{1}{2} \left.\frac{\partial^2 E(k)}{\partial k^2}\right|_{k=0} k^2 + \dots \qquad \dots(1.1)$$

When the band minimum occurs at k = 0, the gradient of E(k) is zero at k = 0, so, the lowest order,

$$E(k) = E(0) + \frac{h^2 k^2}{2m^*} \qquad \dots(1.2)$$

where

$$\frac{1}{m^*} = \frac{1}{h^2} \frac{\partial^2 E(k)}{\partial k^2} \qquad \dots(1.3)$$

The effective mass, and $h = \dfrac{h}{(2\pi)} = 1.05457168\,(18) \cdot 10^{-34}$ Js is Planck's constant.

For diamond crystals such as silicon the conduction band has three minima, one at k = 0 (called the T point), another along (111) directions at the boundary of the first Brillouin zone (called L) and a third one near the zone boundary along (100) directions. cf. If the first conduction band minimum E(k) is described by,

$$E(k) = \frac{h^2 k^2}{2m^*} \qquad \dots(1.4)$$

The constant energy surface in k–space of the band approximation, forms a sphere and the effective mass $m^*$ is isotropic. When the conduction band does not lie at k= 0, the effective mass depends on the crystallographic orientation of the minimum. In common cubic semiconductors, we find that

$$E(k) = \frac{h}{2}\left(\frac{k_l^2}{m_l^*} + \frac{k_t^2}{m_t^*} + \frac{k_t^2}{m_t^*}\right) \qquad \dots(1.5)$$

Equation (1.5) expresses a band with ellipsoidal constant energy surfaces. The effective mass is a different longitudinal and transverse effective masses, $m_l^*$ and $m_t^*$, respectively. Material specific values for $m_l^*$ and $m_t^*$. Formula (1.5) is referred to as the parabolic energy band approximation.

In this case, high applied fields, carriers may be far above the minimum, and the higher order terms in the Taylor series expansion cannot be ignored. For the conduction band, this is approximated by a relation of the form

$$E(1 + \alpha E) = \frac{h^2 k^2}{2m^*} \qquad \dots(1.6)$$

where $m^*$ is determined from formula (1.2) at the minimum. For a minimum at k = 0,

$$\alpha_T = \frac{1}{E_T} = \left(1 - \frac{M_T^*}{m_0}\right)^2 \qquad \dots(1.7)$$

where $E_T$ is the direct bandgap. Formula (1.6) is referred to as the non–parabolic energy band approximation.

### 1.2.1 Full Band Approach

- According to the higher energies the conduction band approximates with non–parabolic parameter $\alpha$ as in equation (1.6). Nevertheless a non–parabolic band provides a reasonable approximation only up to an energy of about 1 eV.

- For example in silicon MOSFETs an important reliability problem is caused by injection of electrons from the channel into the gate oxide. The energy barrier at the $SiO_2$, Si interface is 3.1 eV. For these problems simple expressions for E(k) are not valid any more and a numerically generated table of E(k) must be used.

- A very useful approach to a numerical evaluation of E(k) is the so called pseudo potential method. To evaluate E(k) the Schrödinger equation for the electrons is solved for a bulk semiconductor in the absence of scattering and without any built–in potential. The pseudo potential method relies on the fact that the band structure is largely determined by the valence electrons. In addition empirical form factors have been derived to fit band gaps at the high symmetry locations.

## 1.3 HISTORICAL DEVELOPMENT AND BASIC CONCEPTS OF CRYSTAL STRUCTURE AND DEFECTS

- Many solids and some crystalline liquids have a consistent, repeating, three–dimensional arrangement of atoms referred to as a crystal lattice or crystal structure. In contrast, an amorphous solid is a kind of solid material, for instance glass, which deficiencies such a long–range repeating structure.

- Many of the physical, optical, and electrical properties of crystalline solids or liquids are thoroughly associated to crystal structure. The repeating units of a crystalline structure are referred to as "cells." These units are made up of small boxes or other three–dimensional shapes.

- Many of these cells are grouped together in a repeating, orderly structure to make up the overall structure.

- The crystal structure of a given crystalline material can affect many of that material's overall properties.
- It is one of the major defining factors affecting the optical properties of the material, for instance.
- The crystal structure affects the reactivity of the crystallize material. It regulates the reactive atoms on the outside edges and faces of the crystalline solid or liquid arrangement.
- In other way material traits, as well as electrical and magnetic properties of some materials, are also determined largely by crystal structure. Most of the time atoms self–organize in crystals.
- The crystalline lattice is a periodic array of the atoms. When the solid is not crystalline, it is called amorphous. Examples of crystalline solids are metals, diamond and other precious stones, ice, graphite. Glass, amorphous carbon (aC), amorphous Si, most plastics are common examples of amorphous solids.

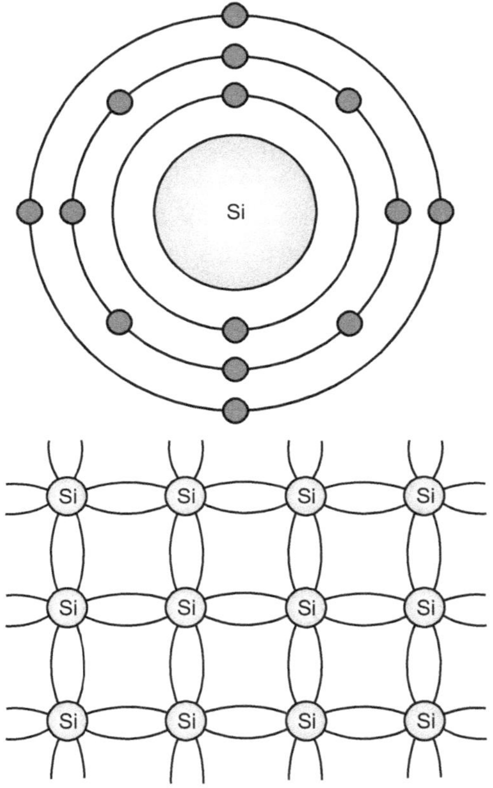


**Fig. 1.3: Silicon atoms**

- The above Fig. 1.3 shows the arrangement of the silicon atoms in a unit cell, with the numbers representing the height of the atom above the base of the cube as a fraction of the cell dimension.
- Silicon crystallines in the same pattern as diamond, in a structure which Ashcroft and Mermin call "two interpenetrating face–centered cubic" primitive lattices.

- The lines between silicon atoms in the lattice illustration specify nearest–neighbor bonds. The cube side for silicon is 0.543 nm. Germanium has the same diamond structure with a cell dimension of 0.566 nm.
- A crystal structure consists of a unit cell, a set of atoms arranged in a particular way which is periodically repeated in three dimensions on a lattice.

**Important Properties of the Unit Cells are:**

- The type of atoms and their radii R.
- Cell dimensions (side in cubic cells, side of base $a$ and height $c$ in HCP) in terms of $R$.
- Number of atoms per unit cell. For an atom that is shared with m adjacent unit cells, we only count a fraction of the atom, 1/m.
- The coordination number, which is the number of closest neighbors to which an atom is bonded.
- The atomic packing factor, which is the fraction of the volume of the cell actually occupied by the hard spheres.

$$APF = \frac{\text{Sum of atomic volumes}}{\text{Volume of cell}}$$

**Table 1.1**

| Unit Cell | n | CN | a/R | APF |
|---|---|---|---|---|
| SC | 1 | 6 | 2 | 0.52 |
| BCC | 2 | 8 | 4Ö 3 | 0.68 |
| FCC | 4 | 12 | 2Ö 2 | 0.74 |
| HCP | 6 | 12 | | 0.74 |

- The closest packed direction in a BCC cell is along the diagonal of the cube; in a FCC cell is along the diagonal of a face of the cube. The unit cell is the smallest structure that repeats itself by translation through the crystal. We construct these symmetrical units with the hard spheres.
- The most common types of unit cells are faced–centered cubic (FCC), Body–Centered Cubic (BCC) and Hexagonal Close–Packed (HCP). Other types exist, particularly among minerals. The Simple Cube (SC) is often used for didactical purpose.

## 1.3.1 Crystal Defects

- The structure of real crystals differs from that of ideal ones. Real crystals always have certain defects or imperfections, and therefore, the arrangement of atoms in the volume of a crystal is far from being perfectly regular.
- Natural crystals always contain defects, often in abundance, due to the uncontrolled conditions under which they were formed. The presence of defects

which affect the colour can make these crystals valuable as gems, as in ruby (chromium replacing a small fraction of the aluminium in aluminium oxide: $Al_2O_3$). Crystal prepared in laboratory will also always contain defects, although considerable control may be exercised over their type, concentration, and distribution.

- The importance of defects depends upon the material, type of defect, and properties, which are being considered. Some properties, such as density and elastic constants, are proportional to the concentration of defects, and so a small defect concentration will have a very small effect on these. Other properties, e.g. the colour of an insulating crystal or the conductivity of a semiconductor crystal, may be much more sensitive to the presence of small number of defects.

- Indeed, while the term defect carries with it the connotation of undesirable qualities, defects are responsible for many of the important properties of materials and much of material science involves the study and engineering of defects so that solids will have desired properties. A defect free, i.e. ideal silicon crystal would be of little use in modern electronics; the use of silicon in electronic devices is dependent upon small concentrations of chemical impurities such as phosphorus and arsenic which give it desired properties. Some simple defects in a lattice are shown in Fig. 1.4.

- There are some properties of materials such as stiffness, density and electrical conductivity which are termed structure–insensitive, are not affected by the presence of defects in crystals while there are many properties of greatest technical importance such as mechanical strength, ductility, crystal growth, magnetic

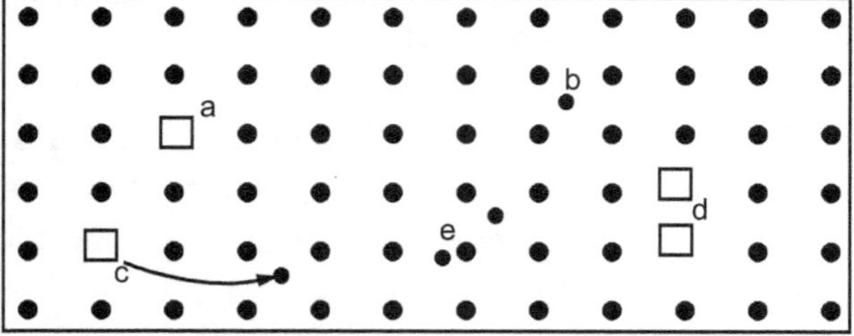

a = vacancy (Schottky defect)
b = interstitial
c = vacancy – interstitial pair
    (Frenkel defect)
d = divacancy
e = split interstitial
▢ = vacant site

**Fig. 1.4 : Some Simple defects in a lattice**

Hysteresis, dielectric strength, condition in semiconductors, which are termed structure sensitive are greatly affected by the–relatively minor changes in crystal structure caused by defects or imperfections. Crystalline defects can be classified on the basis of their geometry as follows:

1. Point imperfections
2. Line imperfections
3. Surface and grain boundary imperfections
4. Volume imperfections

The dimensions of a point defect are close to those of an interatomic space. With linear defects, their length is several orders of magnitude greater than the width. Surface defects have a small depth, while their width and length may be several orders larger. Volume defects (pores and cracks) may have substantial dimensions in all measurements, i.e. at least a few tens of $A^0$. We will discuss only the first three crystalline imperfections.

## 1.3.2 Point Defects in Ionic Crystals and Metals

The point imperfections, which are lattice errors at isolated lattice points, take place due to imperfect packing of atoms during crystallisation. The point imperfections also take place due to vibrations of atoms at high temperatures. Point imperfections are completely local in effect, e.g. a vacant lattice site. Point defects are always present in crystals and their present results in a decrease in the free energy. One can compute the number of defects at equilibrium concentration at a certain temperature as,

$$n = N \exp[-E_d / kT]$$

Where n – number of imperfections, N – number of atomic sites per mole, k – Boltzmann constant, $E_d$ – free energy required to form the defect and T – absolute temperature. E is typically of order of 1 eV since k = $8.62 \times 10^{-5}$ eV /K, at T = 1000 K, n/N = $\exp[-1/(8.62 \times 10^{-5} \times 1000)] \approx 10^{-5}$, or 10 parts per million. For many purposes, this fraction would be intolerably large, although this number may be reduced by slowly cooling the sample.

1. **Vacancies:** The simplest point defect is a vacancy. This refers to an empty (unoccupied) site of a crystal lattice, i.e. a missing atom or vacant atomic site [Fig. 1.5] such defects may arise either from imperfect packing during original crystallization or from thermal vibrations of the atoms at higher temperatures. In the latter case, when the thermal energy due to vibration is increased, there is always an increased probability that individual atoms will jump out of their positions of lowest energy. Each temperature has a corresponding equilibrium concentration of vacancies and interstitial atoms (an interstitial atom is an atom transferred from a site into

an interstitial position). For instance, copper can contain $10^{-13}$ atomic percentage of vacancies at a temperature of 20–25°C and as many as 0.01 % at near the melting point (one vacancy per $10^4$ atoms). For most crystals the–said thermal energy is of the order of I eV per vacancy. The thermal vibrations of atoms increases with the rise in temperature. The vacancies may be single or two or more of them may condense into a di–vacancy or trivacancy. We must note that the atoms surrounding a vacancy tend to be closer together, thereby distorting the lattice planes. At thermal equilibrium, vacancies exist in a certain proportion in a crystal and thereby leading to an increase in randomness of the structure. At higher temperatures, vacancies have a higher concentration and can move from one site to another more frequently. Vacancies are the most important kind of point defects; they accelerate all processes associated with displacements of atoms: diffusion, powder sintering, etc.

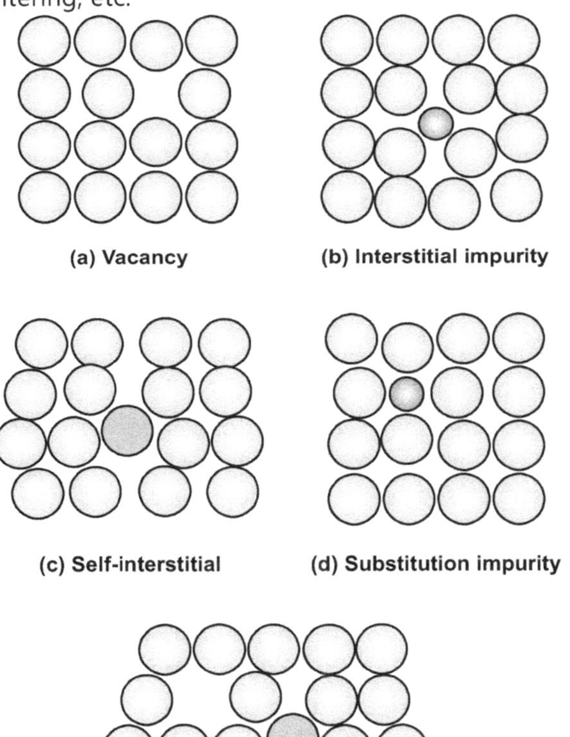

(a) Vacancy      (b) Interstitial impurity

(c) Self-interstitial    (d) Substitution impurity

(e) Frenkel defect

**Fig. 1.5: Point defects in a crystal lattice**

2. **Interstitial Imperfections:** In a closed packed structure of atoms in a crystal if the atomic packing factor is low, an extra atom may be lodged within the crystal structure. This is known as interstitial position,

i.e. voids. An extra atom can enter the interstitial space or void between the regularly positioned atoms only when it is substantially smaller than the parent atoms [Fig. 1.5(b)], otherwise it will produce atomic distortion. The defect caused is known as interstitial defect. In close packed structures, e.g. FCC and HCP, the largest size of an atom that can fit in the interstitial void or space have a radius about 22.5% of the radii of parent atoms. Interstitialcies may also be single interstitial, di–interstitials, and tri–interstitials. We must note that vacancy and interstitialcy are inverse phenomena.

3. **Frenkel Defect:** Whenever a missing atom, which is responsible for vacancy occupies an interstitial site (responsible for interstitial defect) as shown in Fig. 1.5(c), the defect caused is known as Frenkel defect. Obviously, Frenkel defect is a combination of vacancy and interstitial defects. These defects are less in number because energy is required to force an ion into new position. This type of imperfection is more common in ionic crystals, because the positive ions, being smaller in size, get lodged easily in the interstitial positions.

4. **Schottky Defect:** These imperfections are similar to vacancies. This defect is caused, whenever a pair of positive and negative ions is missing from a crystal [Fig. 1.5(e)]. This type of imperfection maintains charge neutrality. Closed–packed structures have fewer interstitialcies and Frenkel defects than vacancies and Schottky defects, as additional energy is required to force the atoms in their new positions.

5. **Substitutional Defect:** Whenever a foreign atom replaces the parent atom of the lattice and thus occupies the position of parent atom (Fig. 1.5(d)], the defect caused is called substitutional defect. In this type of defect, the atom which replaces the parent atom may be of same size or slightly smaller or greater than that of parent atom.

6. **Phonon:** When the temperature is raised, thermal vibrations takes place. This results in the defect of a symmetry and deviation in shape of atoms. This defect has much effect on the magnetic and. electric properties.

All kinds of point defects distort the crystal lattice and have a certain influence on the physical properties. In commercially pure metals, point defects increase the electric resistance and have almost no effect on the mechanical properties. Only at high concentrations of defects in irradiated metals, the ductility and other properties are reduced noticeably.

In addition to point defects created by thermal fluctuations, point defects may also be created by other means. One method of producing an excess number of point defects at a given temperature is by quenching (quick cooling) from a higher temperature. Another method of creating excess defects is by severe deformation of the crystal lattice, e.g., by hammering or rolling. We must note that the lattice still retains its general crystalline nature, numerous defects are introduced. There is also a method of creating excess point defects is by external bombardment by atoms or high–energy particles, e.g. from the beam of the cyclotron or the neutrons in a nuclear reactor. The first particle collides with the lattice atoms and displaces them, thereby causing a point defect. The. number of point defects created in this manner depends only upon the nature of the crystal and on the bombarding particles and not on the temperature.

### 1.3.3 Crystal Growth and Wafer Fabrication

The preparation of silicon or gallium arsenide wafer is required before the fabrication of the integrated circuit.

**The Preparation of Wafer Involves 4 steps.**

1. Distillation and reduction/synthesis,

2. Crystal growth,

3. Grind/saw/polish, and

4. Electrical and mechanical characterizations.

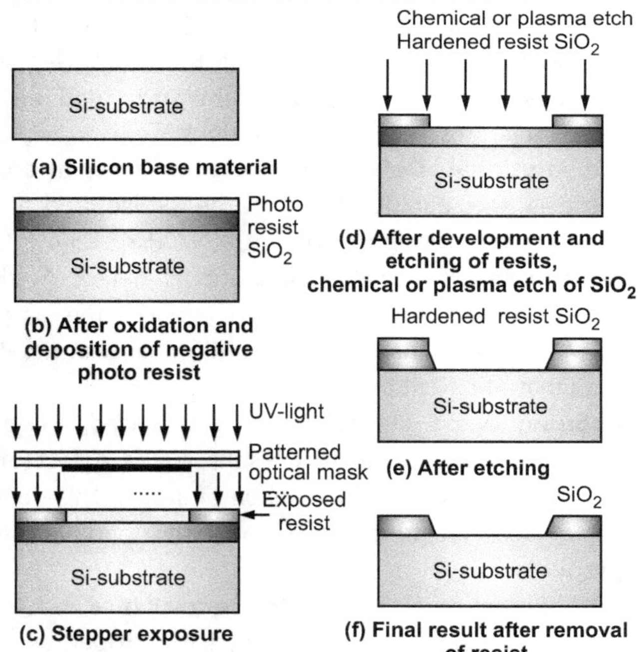



**Fig. 1.6: Crystal growth and wafer fabrication**

- The starting material is silicon dioxide for making silicon wafer. It is chemically processed to form a high–purity crystal polycrystalline semiconductor for which single crystal is formed.

- The single crystal is shaped to define diameter and is sawed into wafer. The wafer is then etched and polished to provide smooth surface. Pure form of sand $SiO_2$ called quartzite is placed in high temperature furnace with various forms of carbon like coke, coal, and even wood chip. Owing silicon dioxide is very stable, carbon is used to replace silicon to form carbon dioxide at reduced temperature. Although there are numbers of reaction take place and the overall reaction follows equation

$$SiO_2 + 2C \rightarrow Si + 2CO \uparrow \qquad \ldots(1.8)$$

- This process generates polycrystalline silicon with about 98% to 99% purity called crude silicon or metallurgical–grade silicon MGS.

- In next process step is the silicon purification step. Silicon is pulverized and treated with hydrochloric acid gas HCl at temperature 300°C to form trichlorosilane $SiHCl_3$ vapor. The chemical reaction follows equation

$$Si + 3HCl \xrightarrow{300°C} SiHCl_3 + H_2 \uparrow \qquad \ldots(1.9)$$

- Trichlorosilane TCS vapor is then gone to fractional distillation to remove unwanted impurities through a series of filters, condensers (boiling point 32°C), and purifiers to finally get an ultra high purity liquid of purity higher than 99.9999999% at room temperature. Then, high–purity TCS is used in the hydrogen reduction reaction at temperature 1,1000C to produce the electronic grade silicon EGS.

$$SiHCl_3 + H_2 \xrightarrow{1100°C} Si + 3HCl \uparrow \qquad \ldots(1.10)$$

- The reaction takes place in a reactor containing resistance heated silicon rod, which serves as the nucleation point for deposition of EGS in polycrystalline form of high purity. This is the raw material used to prepare device quality single crystal. Pure EGS has impurity concentration generally in part per billion. The pure EGS is then ready to be pulled into silicon ingot for making wafer for integrated circuit fabrication. There are a number of methods used to grow silicon crystalline ingot.

### Silicon Wafer Fabrication Process

- More than 90% of the earth's crust is composed of Silica ($SiO_2$) or Silicate, making silicon the second most abundant element on earth. When sand glitters in sunlight, that's silica.

- Silicon is found in compounds in nature. Silicon is the principle platform for semiconductor devices.

- Now–a–days semiconductor is the most advanced technology requires the monocrystalline silicon with uniform chemical characteristics to control doping and oxygen content.

- The process to transform raw silicon into a useable single–crystal substrate for modern semiconductor processes begins by mining for relatively pure Silicon Dioxide.

- Most silicon now is made by reducing temperature of $SiO_2$ with Carbon in an electric furnace from 1500 to 200 °C. With carefully selected pure sand, the result is commercial brown Metallurgical Grade Silicon of 97% purity or better.

- This is the silicon eventually used for semiconductors, but it must be further purified to bring impurities below the parts–per–billion level. Now that a high level of purity has been attained (99.999999999%), the atomic structure of the silicon must be dealt with a process termed as Crystal Growing which transforms polycrystalline silicon into samples with a singular crystal orientation, known as ingots.

- The Polysilicon is mechanically broken into 1 to 3 inch chunks and undergoes stringent surface etching and cleaning in a clean room environment. These chunks are then packed into quartz crucibles for meltdown (at 1420 °C) in a CZ furnace.

- A mono crystalline Silicon seed is installed into a seed shaft in the upper chamber of the furnace. Slowly, the seed is lowered so that it dips approximately 2 mm into the Silicon melt.

- Next, the seed is slowly retracted from the surface allowing the melt to solidify at the boundary.

- As the seed pulls the Silicon from the melt, both the crucible and the seed are rotated in opposite directions to allow for an almost round crystal to form. CZ furnaces also must be very stable and isolated from vibrations.

- The crystal diameter is properly achieved by increasing the seed lift along with the heat transfer from heating elements it will control the diameter of the crystal. In this growth process, the crucible slowly dissolves oxygen and melt that into the final crystal.

- Additionally, to pull speed and heat transfer at the solid–liquid interface, heat dissipation during crystal cooling strongly determines microscopic defect characteristics in the final crystal. For modern CZ pulling systems, those variables can be accurately predicted by numerical simulations which allow designing the geometrical and thermal configuration of the CZ puller to the desired outcome of the crystals. Once the growth process is complete, the crystal is cooled inside the furnace for up to 7 hours. This gradual cooling allows the crystal lattice to stabilize and makes handling easier before transport to the next operation.

- It is important, for some applications, to have even lower concentrations of impurity atoms (e.g. Oxygen) than what can be achieved by CZ crystal growth. In this case, Float Zone Crystal Growth is used.

- In this process the end of a long polysilicon rod is locally melted and brought in contact with a monocrystalline Silicon seed. The melted zone slowly migrates through the poly rod leaving behind a final uniform crystal. Ingots coming from crystal growing are slightly over–sized in diameter and typically not round.

- Hence, a machine employing a grind wheel shapes the ingot to the precision needed for wafer diameter control. Other grinding wheels are then used to carve a characteristic notch or a flat in order to define the proper orientation of the future wafer versus a particular crystallographic axis.

- Wafer shaping contains a series of precise mechanical and chemical process steps that are necessary to turn the ingot segment into a functional wafer.

- It is during these steps that the wafer surfaces and dimensions are perfected to exacting detail. Each step is designed to bring the wafer into compliance with each customer specification.

- The first of these critical steps is Multi–Wiring Slicing. The dominant state of the art slicing technology refers to Multi–Wire Sawing (MWS). Here, a thin wire is arranged over cylindrical spools so that hundreds of parallel wire segments simultaneously travel through the ingot.

- While the saw as a whole slowly moves through the ingot, the individual wire segments conduct a translational motion always bringing fresh wire into contact with the Silicon.

- The sawing effect is actually achieved by SiC or other grinding agents that run along the rotating wire.

- After MWS the wafers are cleaned and consolidated into process lots and transported to the next operation.

- The sideward deflection of the wire saw can lead to marks or "waviness" on the wafer surface and wire–to–wire thickness variations cause wafer thickness variations of up to several microns. Thus wafers are exposed to a complex polishing process.

## 1.3.4 Crystal Planes and Orientation

The orientation of a plane in a lattice is stated by Miller indices. They referred as intercept of the plane with the axes along the primitive translation vectors $A_1$, $A_2$ and $A_3$. Let us these intercepts be X, Y, and Z, so that X is fractional multiple of $A_1$, Y is a fractional multiple of $A_2$ and Z is a fractional multiple of $A_3$. Therefore we can measure X, Y, and Z in units $A_1$, $A_2$ and $A_3$ respectively. We have a triplet of integers (X Y Z) then we invert it (1/X 1/Y 1/Z) and reduce this set to a similar one having the smallest integers by multiplying by a common factor. This set is called Miller indices of the plane (hkl).

For example, if the plane intercepts X, Y, and Z in points 1, 3, and 1, the index of this plane will be (313) as shown in Fig. 1.7.

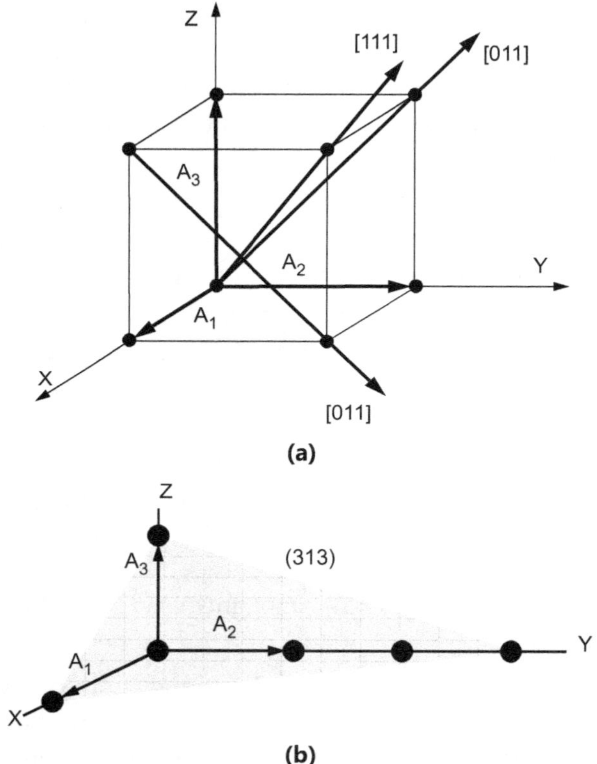


**Fig. 1.7: Crystal plane intercept**

The orientation of a crystal plane is determined by three points in the plane, provided they are not collinear. If each point lay on a different crystal axis, the plane could be specified by giving the coordinates of the points in terms of the lattice constants a, b, c. A notation conventionally used to describe lattice points (sites), directions and planes is termed as Miller Indices. A crystal lattice may be considered as an assembly of equidistant parallel planes passing through the lattice points and are known as lattice planes. In order to specify the orientation one employs the so called Miller indices. For simplicity, let us start with a two dimensional lattice and then generalized to three dimensional case.

The equation of plane in 2D and 3D having the intercepts a, b and a, b, c respectively are:

$$\frac{x}{a} + \frac{y}{b} = 1 \text{ and } \frac{x}{a} + \frac{y}{b} + \frac{z}{c} = 1$$

Crystal direction is the direction (line) of axes or line from the origin and denoted as [111], [100], [010] etc.

## 1.4 MODERN CMOS TECHNOLOGY

- A Complementary Metal Oxide Semiconductor (CMOS) is a kind of integrated circuit technology. The term is frequently used to refer to a battery–powered chip found in many PCs that holds some basic information, as well as the date and time and system configuration settings, required by the Basic Input/Output System (BIOS) to start the computer. However, this name is rather deceptive as most modern computers no longer use these chips for this function; on the other hand they depend on other types of non–volatile memory. CMOS chips are still found in many other electronic devices, including digital cameras.

- In addition to NMOS and PMOS transistors, the technology offers:
  - ➢ A deep n–well that can be utilized to reduce substrate noise coupling.
  - ➢ A MOS varactor that can serve in VCOs
  - ➢ At least six levels of metal that can form many useful structures such as inductors, capacitors, and transmission lines.

**Why CMOS Technology?**

Comparison of BJT and MOSFET technology from an analog viewpoint as follows:

| Feature | BJT | MOSFET |
|---|---|---|
| Cutoff Frequency $(f_T)$ | 100 GHz | 50 GHz (0.25 m) |
| Noise (thermal about the same) | Less 1/f | More 1/f |
| DC Range of operation | 9 decades of exponential current versus vBE | 2–3 decades of square law behavior |
| Small Signal Output Resistance | Slightly larger | Smaller for short channel. |
| Switch Implementation | Poor | Good |
| Capacitor Implementation | Voltage dependent | Reasonably good |

- Almost every comparison favors the BJT, however a similar comparison made from a digital viewpoint would come up on the side of CMOS.

- Therefore, since large–volume technology will be driven by digital demands, CMOS is an obvious result as the technology of availability.

- The potential for technology improvement for CMOS is greater than for BJT.

- Performance generally increases with decreasing channel length.

**Basic Steps of CMOS Technology**

1. Oxide growth
2. Thermal diffusion
3. Ion implantation
4. Deposition
5. Etching
6. Epitaxial

- Photolithography is the means by which the above steps are applied to selected areas of the silicon wafer as shown in following Fig. 1.8.

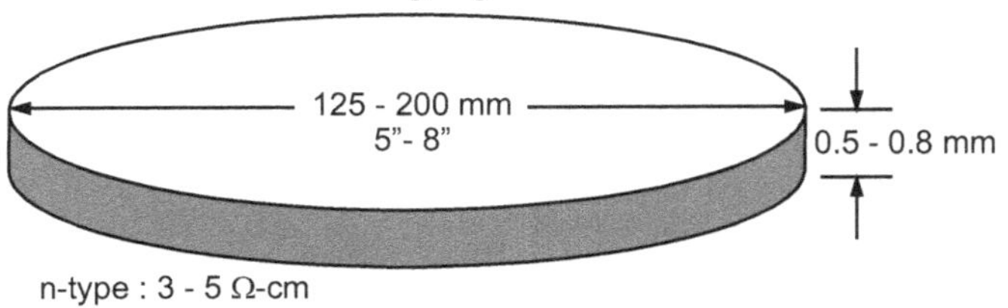


**Fig. 1.8: Si wafer**

## 1. Oxidation

Oxidation is the process by which a layer of silicon dioxide is grown on the surface of a silicon wafer.

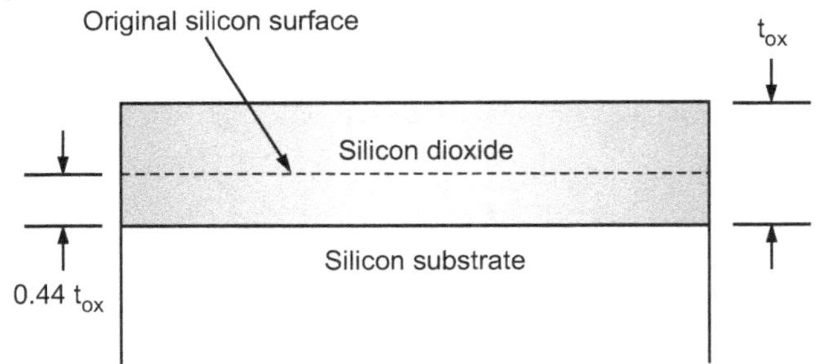


**Fig. 1.9: Oxidation**

**Uses**

- Protect the underlying material from contamination.

- Provide isolation between two layers.

Very thin oxides (100 °A to 1000 °A) are grown using dry oxidation techniques. Thicker oxides (>1000 °A) are grown using wet oxidation techniques.

## 2. Diffusion

The process of movement of impurity atoms at the surface of the silicon into the bulk of the silicon. It is always in the direction from higher concentration to lower concentration.

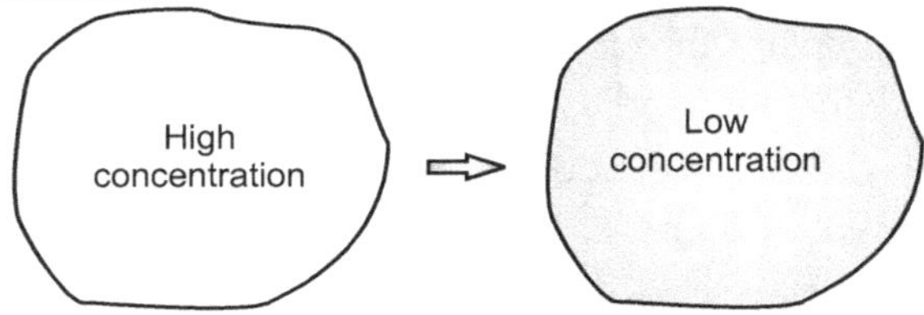


**Fig. 1.10: Diffusion**

It is typically done at high temperatures: 800 to 1400 °C

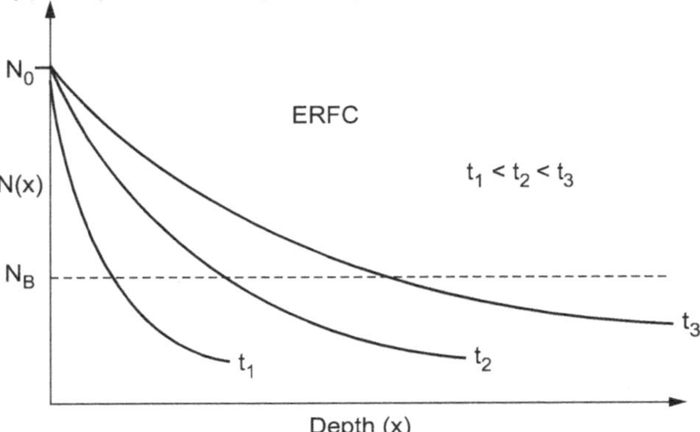


**Fig. 1.11 : Infinite source of impurities at the surface**

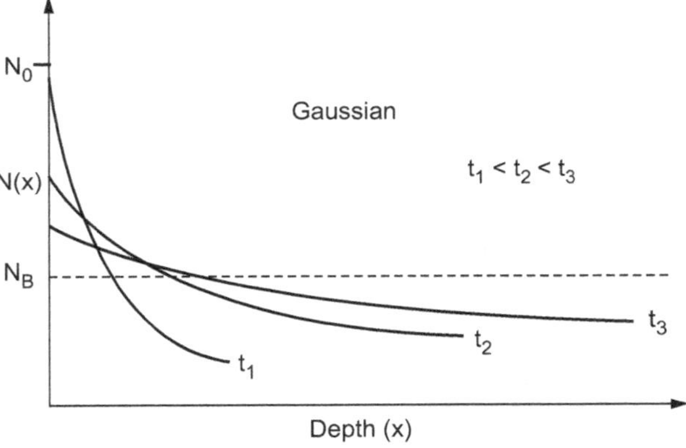


**Fig. 1.12 : Finite source of impurities at the surface**

## 3. Ion Implantation

The process by which impurity ions are accelerated to a high velocity and physically lodged into the target material.

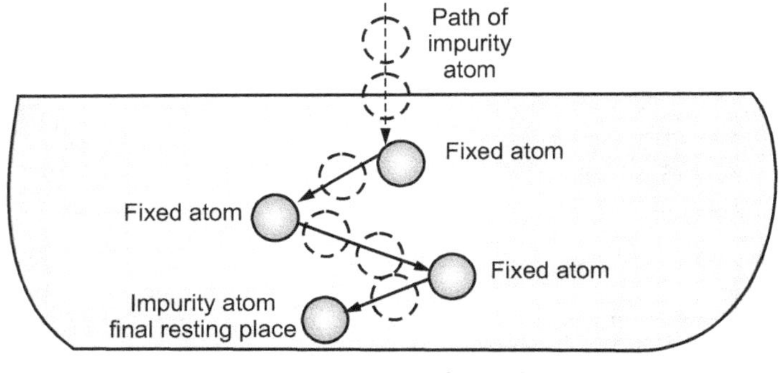


**Fig. 1.13: Ion implantation**

- Annealing is required to activate the impurity atoms and repair the physical damage to the crystal lattice. This step is done at 500 to 800°C.
- Ion implantation is a lower temperature process compared to diffusion.
- Can implant through surface layers, thus it is useful for field–threshold adjustment.
- Can achieve unique doping profile such as buried concentration peak.

### 4. Deposition

- The deposition means by which various materials are deposited on the silicon wafer.

**Examples:**

- Silicon nitride ($Si_3N_4$)
- Silicon dioxide ($SiO_2$)
- Aluminum
- Polysilicon
- There are various ways to deposit a material on a substrate:
  - ➤ Chemical–Vapor Deposition (CVD)
  - ➤ Low–Pressure Chemical–Vapor Deposition (LPCVD)
  - ➤ Plasma–Assisted Chemical–Vapor Deposition (PACVD)
  - ➤ Sputter deposition material that is being deposited using these techniques covers the entire wafer.

### 5. Etching

- The Etching process of selectively removing a layer of material. When etching is performed, the etchant may remove portions or all of:
  - ➤ The desired material.
  - ➤ The underlying layer.
  - ➤ The masking layer.

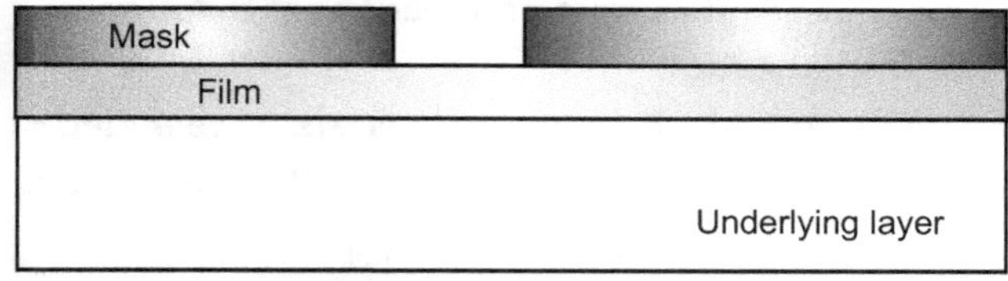


(a) Portion of the top layer ready for etching.

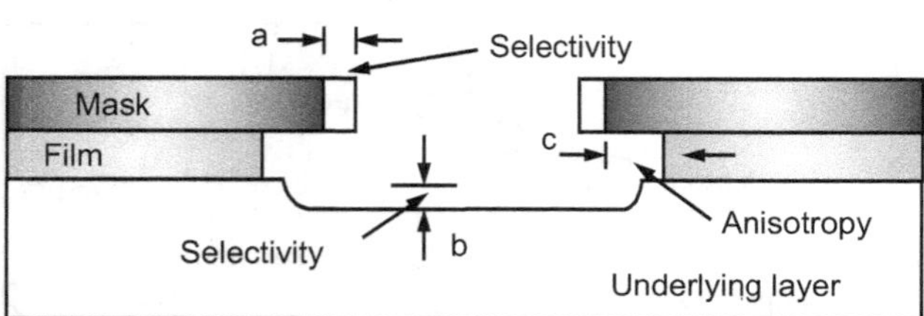


(b) Horizontal etching and etching of underlying layer.

**Fig. 1.14: Etching**

**Important Considerations:**

- Anisotropy of the etch is defined as, A = 1–(lateral etch rate/vertical etch rate)
- Selectivity of the etch (film to mask and film to substrate) is defined as,

$$S_{film-mask} = \frac{film\ etch\ rate}{mask\ etch\ rate}$$

$$A = 1\ and\ S_{film-mask} = \infty\ are\ desired$$

There are basically two types of etches:

1. Wet etch which uses chemicals
2. Dry etch which uses chemically active ionized gases.

### 6. Epitaxial

Epitarial of growth is the formation of a layer or of single crystal silicon on the surface of the silicon material so that the crystal. Structure of the silicon is continuous across the following interfaces.

- It is done externally to the material as opposed to diffusion which is internal.
- The epitaxial layer (epi) can be doped differently, even opposite, of the material on which it grown.
- It accomplished at high temperatures using a chemical reaction at the surface.
- The epi layer can be any thickness, typically 1–20 microns.

**Fig. 1.15: Epitaxial Layer**

### Basic CMOS Concepts

A transistor behaves like a switch. For NMOS transistors, if the input is a 1 the switch is on, otherwise it is off. Conversely, for the PMOS, if the input is 0 the transistor is on, then the transistor is off. Here, it is a graphical representation of these facts:

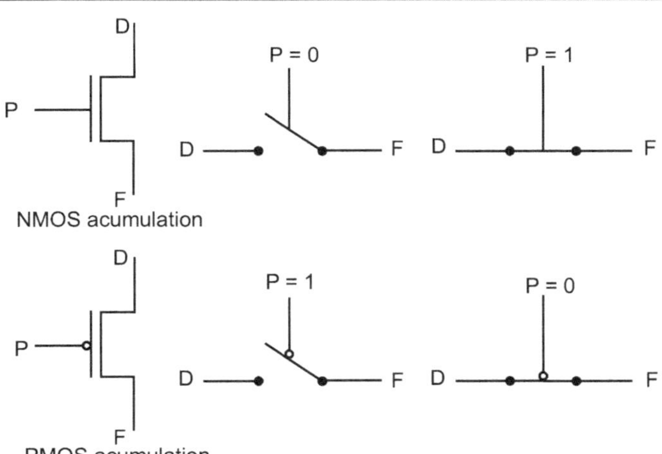



**Fig. 1.16: Accumulation of NMOS and PMOS**

- When a circuit contains both NMOS and PMOS transistors, we say it is implemented in CMOS (Complementary MOS) Understanding the basics of transistors, we can now design a simple NOR gate.

- Following Fig. 1.17 shows the implementation in transistors of the NOT gate and how it works for different inputs (1 and 0). On the left there is the implementation, on the right the behavior. The symbol VDD is the source voltage (or the logic 1), GND is the ground (or the logical 0).

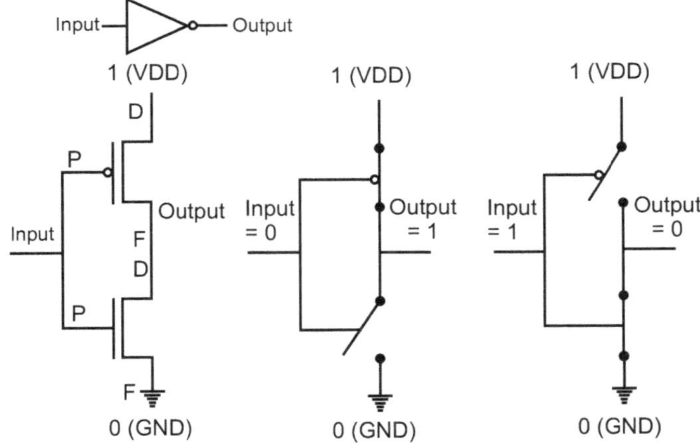



**Fig. 1.17: NOT operation**

- We have just seen how to implement a simple logic gate using transistors. To implement the rest of logical gates (and whatever circuit we might think off), we will analyze first the behavior of the transistors when connected in a "series" fashion or in a "parallel" way.

- If we connect two NMOS transistors in series, we get the behavior shown in following Fig. 1.18. It is triangle in the bottom representation of ground (GND).

**Output = 0, Si S1 = 1, S2 = 1**

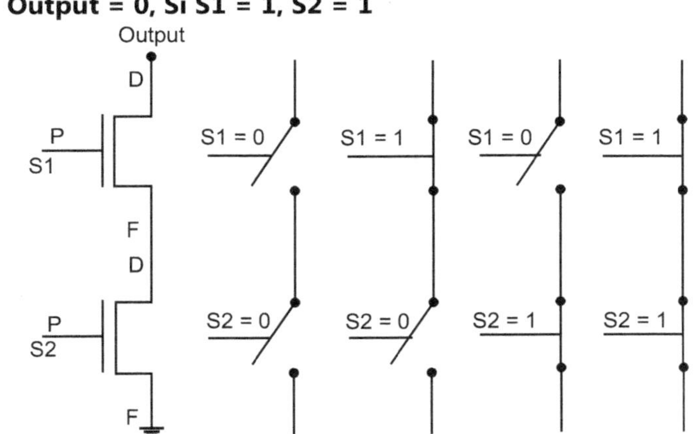



**Fig. 1.18: Series connection of NMOS transistor**

- Following Fig. 1.18 shows the behaviour of the PMOS when connected in series. It is horizontal line on the top of the first transistor representation of $V_{DD}$.

**Output = 1 Si S1 = 0 ; S2 = 0**

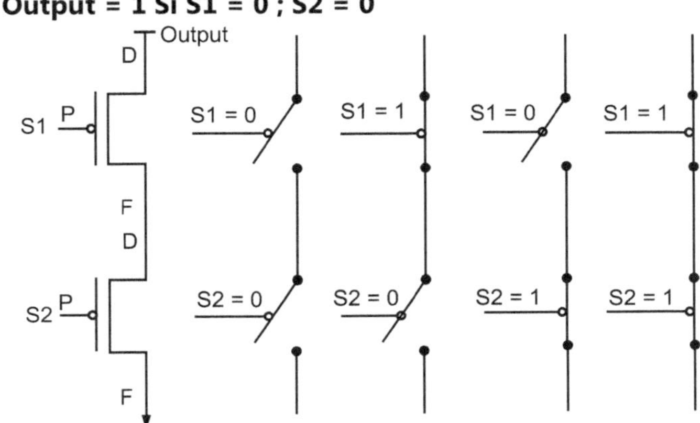



**Fig. 1.19: Series connection of PMOS transistor**

- When using CMOS technology, precisely static CMOS, we will design the circuits with two clearly defined parts. One (called pull–up) will be built of PMOS transistors and it has the duty of setting the output to 1 whenever the implemented function defines it.

- The other part (called pull–down) will be built of NMOS transistors and it will set the output to 0 whenever the implemented function defines it. All circuits will either set the output to 1 or 0 for any combination of the input values. Both pull–up and pull–down cannot be active at the same time (when the output set 1 and input set 0 at the same time). Similarly, both the pull–up and the pull–down cannot be off at the same time (logic functions have always a defined output either 0 or 1).

- Nevertheless, we will see further down the course that when not implementing logic functions we might be interested sometimes in setting the output to undetermined in certain cases.

### Complementary CMOS Logic Gates

- nMOS pull–down network
- pMOS pull–up network and static CMOS

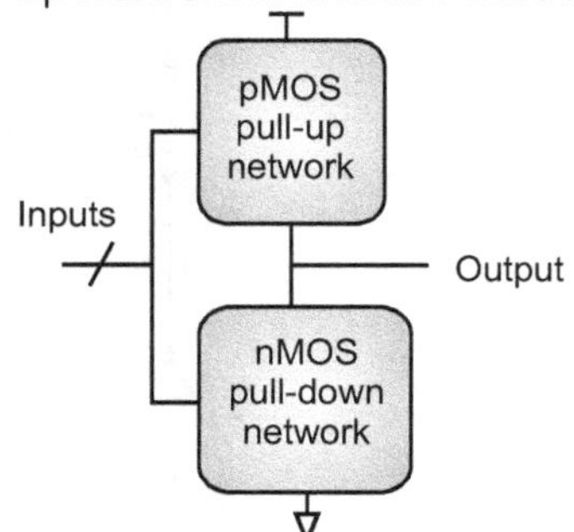


**Fig. 1.20: Pull–down network of CMOS**

|  | Pull–up OFF | Pull–up ON |
|---|---|---|
| Pull–down OFF | Z (float) | 1 |
| Pull–down ON | 0 | X (crossbar) |

As shown in following Figures Fig. 1.21 and Fig. 1.22 demonstrate the implementations of the NAND and NOR gates in CMOS. For each one of them, there is the truth table and clear indications of what outputs are set by the pull–up and what outputs for the pull–down.

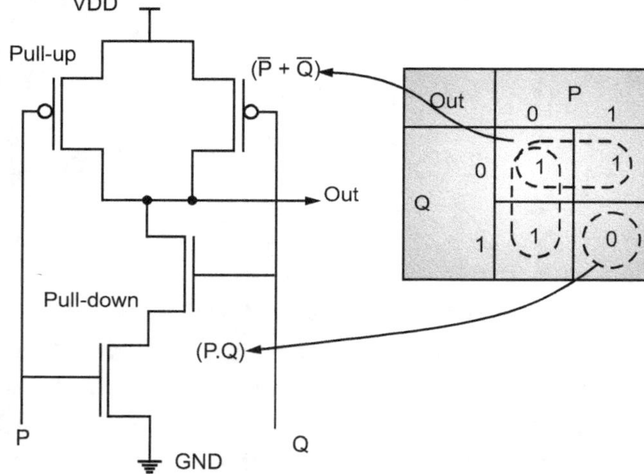


**Fig. 1.21 : NAND gate**

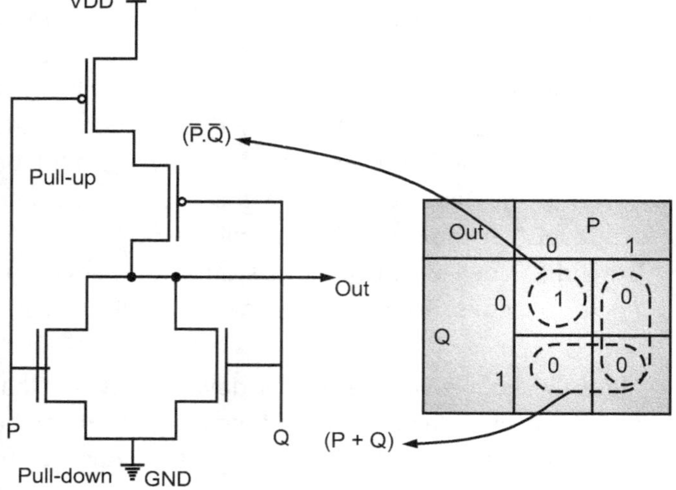


**Fig. 1.22 : NOR gate**

In the same way we implement the logic gates, or we can implement the logic function. Following Fig. 1.23 shows the implementation of the logic function:

$$f(P, Q, R, S) = \overline{S \cdot (P + Q + R)}$$

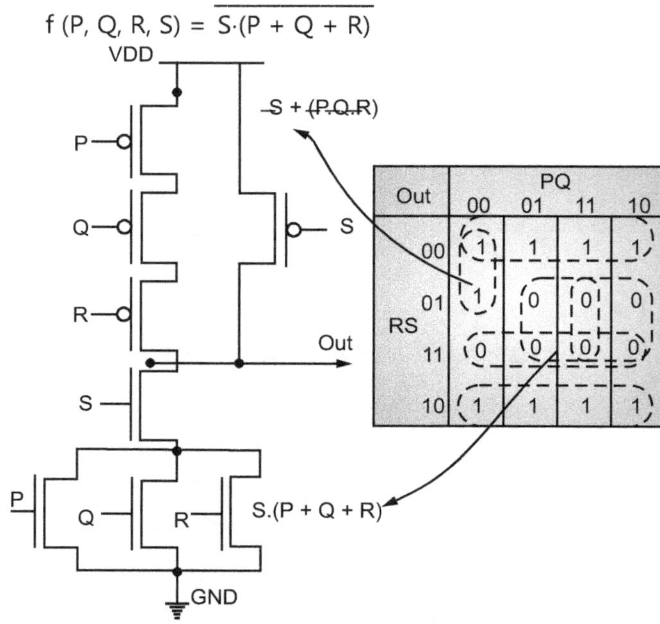


**Fig. 1.23: Implementaion of logical function**

We can design methodology for turning logical functions into CMOS circuits:

- The logic function must be complemented. (i.e., it must look like f(x,y,z) = NOT (expression); in a logic expression,

$$f(x, y, z) = \overline{expression}$$

- AND operator ("·"):
- Pull–down: NMOS transistors NMOS in series
- Pull–up: PMOS transistors in parallel
- OR operator ("+"):
- Pull–down: NOMS transistors in parallel
- Pull–up: PMOS transistors in series

### 1.4.1 Construction of MOS Field Effect Transistor

- The term MOSFET is an acronym for Metal Oxide Semiconductor Field Effect Transistor, and the name indicates a hint to its manufacturer. The devices had been known about for several years but only became important in mid and late 1960s.

- Primarily, semiconductor research had focused in developing the bipolar transistor, and problems had been experienced in fabricating MOSFETs because of process problems, particularly with the insulating oxide layers.

- Now the technology is one of the most widely used semiconductor techniques, having become one of the principle elements in integrated circuit technology today.

- Their performance has enabled power consumptions in ICs to be reduced. This has reduced amount of heat being dissipated and enabled the large ICs, we take for granted today to become a reality. As a result of this the MOSFET is the most widely used form of transistor in existence today.

- MOSFET have metal gates which are insulated from the semiconductor by a layer of $SiO_2$ or dielectric. In EMOSFET of gate voltage activates the channel by inducing carriers layer between the source and drain under the terminal.

- In depletion type MOSFETs, there is a small strip of semiconductor of the same type as that of the source and drain, and the gate voltage can either reduce (by depleting carriers) or increase (by increasing carriers) the channel current. In an n channel MOSFET, the conducting channel exists in a p type substrate.

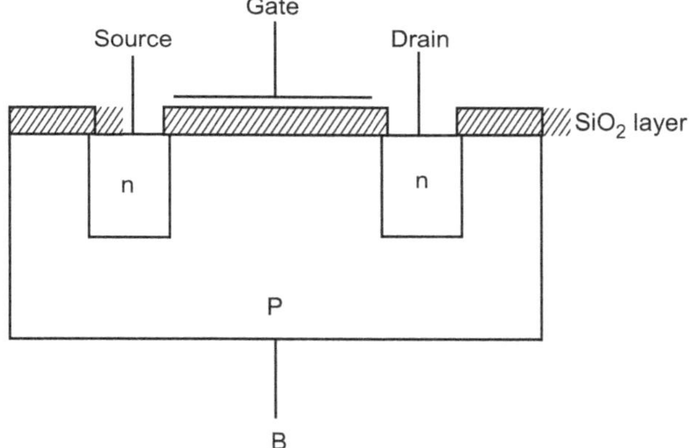


**Fig. 1.24 : *n*–channel EMOSFET structure**

**Note :** The additional B terminal on the substrate, which is often connected directly to the source.

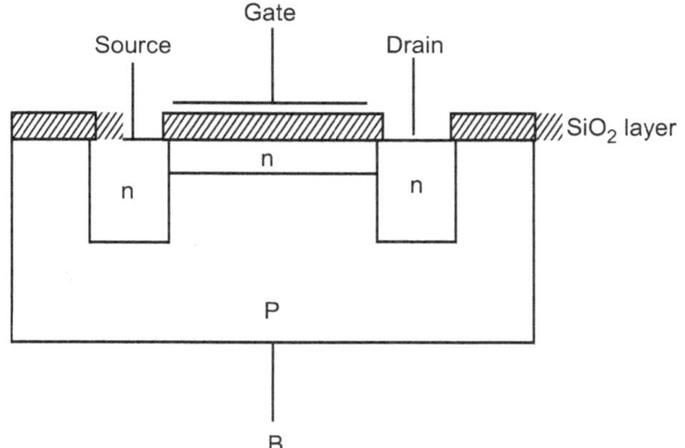


**Fig. 1.25 : n–channel DMOSFET structure**

## Working Principle of MOSFET

- MOSFET depends on the MOS capacitor; the main part of MOSFET is the semiconductor surface at below the oxide layer and between the drain and source terminal can be inverted from p–type to n–type by applying a positive or negative gate voltages separately.

- When applying positive gate voltage the holes present beneath the oxide layer repulsive force are pushed downward with the substrate. The depletion region is populated by the bound negative charges, which are associated with the acceptor atoms.

- The positive voltage also attracts to electrons from the n+ source and drain regions into the channel. The electron reach channel is formed.

- If a voltage is applied between the source and the drain, current flows freely between the source and drain gate voltage controls the electrons concentration the channel becomes positive, if we apply negative voltage, a hole channel will be formed beneath the oxide layer. The controlling of source to gate voltage is responsible for the conduction of current between source and the drain. If the gate voltage exceeds a given value, called the three voltage only the conduction begins.

- The current equation of MOSFET in triode region is :

$$I_D = u_n\, C_{ax}\, \frac{W}{2}\left[(V_{GS} - V_{TH})\, V_{DS} - \frac{1}{2} V_{DS}^2\right]$$

Where, $u_n$ = Mobility of the electrons $C_{ox}$ = Capacitance of the oxide layer W = Width of the gate area L = Length of the channel $V_{GS}$ = Gate to Source voltage $V_{TH}$ = Threshold voltage $V_{DS}$ = Drain to Source voltage.

## P–Channel MOSFET

- MOSFET which has p – channel region between sources and gate is called as p – channel MOSFET. Terminals are gate, drain, source and substrate or body. The drain and source is heavily doped p+ region and the substrate is in n–type.

- The current flows through positively charged holes that's why it is called as p–channel MOSFET.

- If we apply negative gate voltage to the gate terminal of MOSFET, the electrons present beneath of the oxide layer, then repulsive force is pushed downward into the substrate. The depletion region of this case populated by the bound positive charges it has associated with donor atoms.

- The negative gate voltage also attracts holes from p+ source and drain region in to the channel region, thus holes which channel is formed now if a voltage between the source and the drain is applied current flows. The gate voltage controls the hole concentration of the channel.

- The following Fig. 1.26 showing of p–channel enhancement and depletion MOSFET are given below.

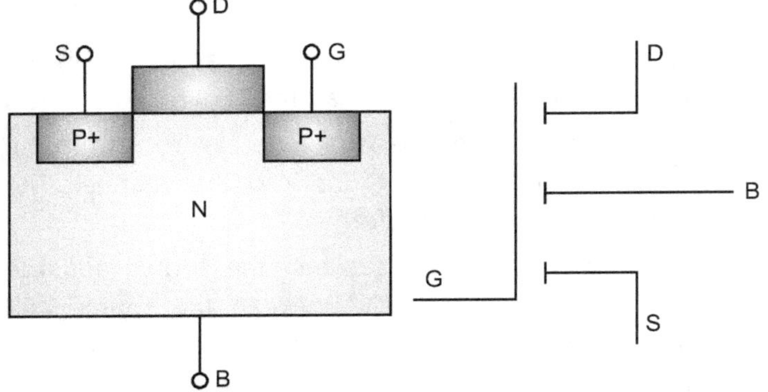

**Fig. 1.26 : Enhancement mode**

**Fig. 1.27 : Depletion mode**

## N–Channel MOSFET

- MOSFET having n–channel region between source and drain is called as n–channel MOSFET.

- The terminals are gate, drain and source and substrate or body. The drain and source is heavily doped n+ region and the substrate is p–type. The current flows because of flow of the negatively charged electrons, that's why it is called as n– channel MOSFET.

- If we apply the positive gate voltage to the gate terminal the holes present beneath the oxide layer, repulsive force (holes) are pushed downwards into the bound negative charge associated with the acceptor atoms.

- The positive gate voltage also attracts to electrons from n+ source and drain region into the channel, thus an electron which channel is formed, now if a voltage is applied between the source and drain.

- The gate voltage controls the electron concentration in the channel n–channel MOSFET is preferred over p–channel MOSFET as the mobility of electrons are higher than holes. The diagrams of enhancements mode and depletion mode are given below.

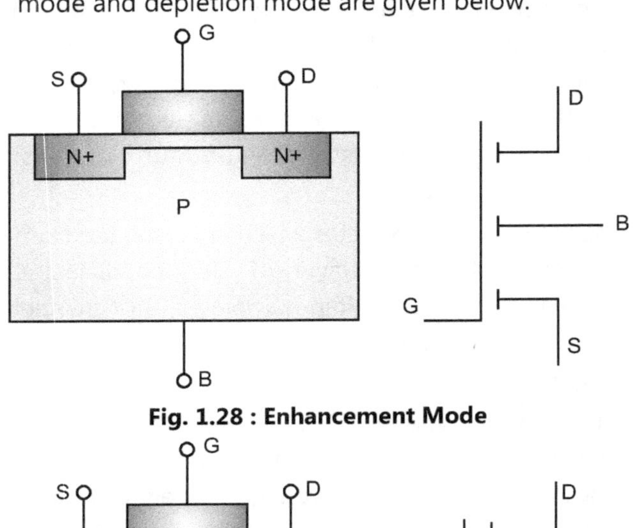

**Fig. 1.28 : Enhancement Mode**

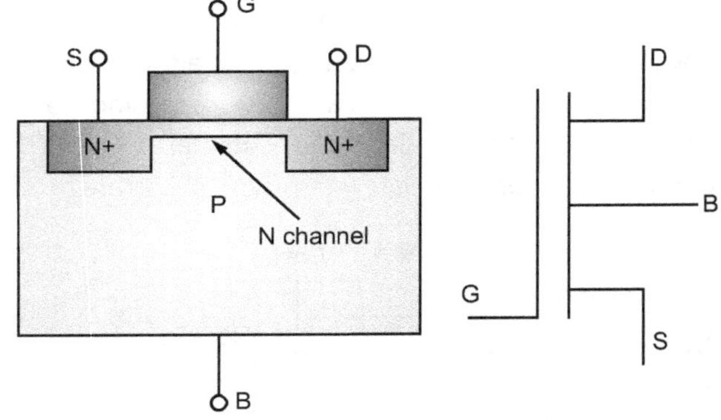

**Fig. 1.29 : Depletion mode**

## 1.5 ELECTRICAL CHARACTERIZATION

- Electrical Characterization can be used to determine resistivity, carrier concentration, mobility, contact resistance, barrier height, depletion width, oxide charge, interface states, carrier lifetimes, and deep level impurities.

- Electrical characterization techniques are used to dielectric–semiconductor edges and to transmit the results of these measurements to the reliability of the device or component. Use this techniques investigators prefer to do not require special test structures. It is used to easily and readily applied to standard test structure to manufactured for the process of file development.

- The continued evolution of semiconductor devices with smaller dimensions to improve performance speed, functionality and integration density, all at reduced cost require layers or films of semiconductors, insulators and metals with increasingly higher quality that are well characterized and can be deposited and patterned to very high precision.

- However, it is not always the case that improvements in materials' quality have kept pace with the evolution of integrated circuit dimensional down–scaling.

- An important aspect of assessing the material quality and device reliability is the development and use of fast, non–destructive, and accurate electrical characterization techniques to determine important parameters such as carrier doping density, type and mobility of carriers, interface quality, oxide trap density. Semiconductor bulk defect density. contact and other parasitic resistances and oxide electrical integrity.

- Electrical Characterization is the key means to characterize and validate your product quality, but from tester to nano–probing, use cases and technologies involved are different: from a simple oscilloscope/ probing through the pads down to most advanced nano–probing solutions/ applications.

- The characterization of materials, devices and circuits is unthinkable without electrical measurements. A broad spectrum of electrical measurement techniques and analysis methods for device characterization.

- This includes capacitance measurements, such as C(V), C(T) and C(f), current measurements with femto–ampere resolution at temperatures between 5 K and 450 K and voltages up to 3000 V.

- Samples can be analyzed by direct probing on wafer level using single probes or probe cards. Package level testing can be performed for long–term reliability characterization. Additionally, carrier lifetime measurements are available on substrates with microwave detected photoconductivity.

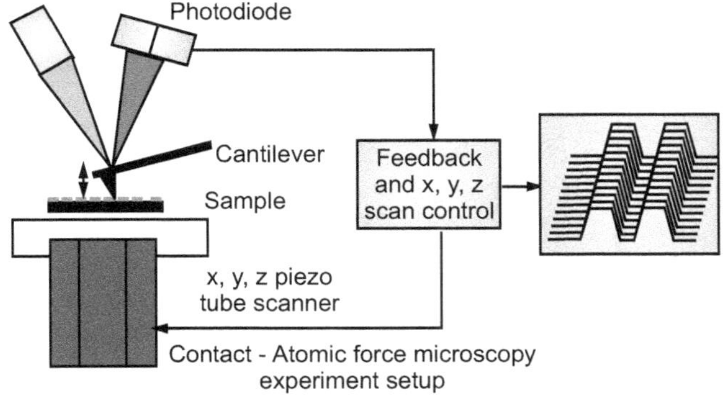


**Fig. 1.30 : Characterization of material**

**The Established Methods Involve:**

- Analysis of single memory cells (memory window, retention, endurance, ...) by static and pulsed measurements.

- Determination of transistor and capacitor characteristic curves by C–V and I–V measurements.

- Determination of sheet resistance for thin layers.

- Determination of doping profiles by Scanning Spreading Resistance Microscopy (SSRM).

- Reliability characterization of dielectric and transistors.

- Defect characterization by charge pumping and charge trapping analysis and defect spectroscopy.

- Measurement of charge carrier mobility with Hall and split–C(V).

Low temperature measurements open the door to investigate effects of charge discreteness or the quantum nature of electrons in devices. In–house fabricated Si–Nanowire transistors operated in the quantum dot regime at T = 5 K show characteristic coulomb oscillations. Here, the periodic peak spacing (21 mV) is directly linked to the specific capacitance and energy structure of the respective nanowire device. Moreover, magneto–transport measurements on transistor devices can be performed up to a magnetic field B= +/– 2.5 T.

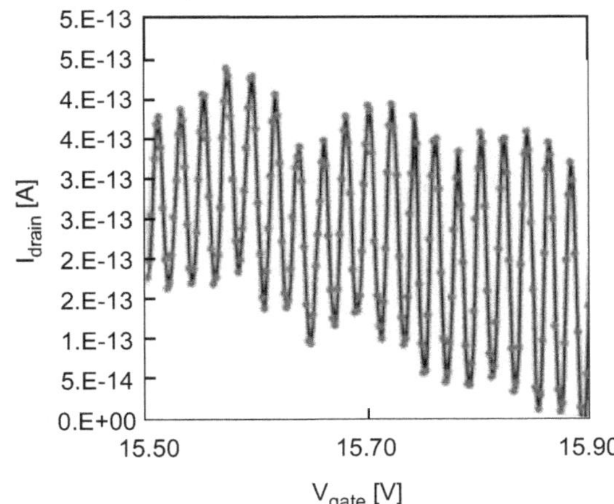


**Fig. 1.31: Low temperature measurement**

A powerful characterization software was developed enabling automatic wafer mapping, easy measurement flow setup in Extensible Markup Language (XML), control of various probe stations and all kinds of measurement equipment, real time parameter extraction and dynamic measurement flow control as well as data export into standard formats (.xlsx, .opj) and user defined formats. The characterization software provides the needed flexibility in handling of all different kinds of samples and the whole band of characterization techniques.

## 1.5.1 Current–Voltage (I–V)/Capacitance–Voltage (C–V) Characterization

- Low current measurement challenges such as error sources for example leakage currents, noise, offset currents, piezoelectric currents, and environmental conditions can have a serious impact on measurement accuracy.

- Precision DC I–V measurements are the foundation of electrical characterization for cutting–edge devices, materials, and semiconductors. With proper measurement techniques and practices, these critical measurement challenges can be met.

- Now–a–days, many parametric measurements require a fast, pulsed I–V measurement using pulsed I–V signals to characterize devices rather than DC signals makes it possible to study or reduce the effects of self–heating (joule heating) or to minimize current drift/degradation in measurements.

- Many applications require pulsed I–V along with C–V and DC I–V measurements. Learn how to combine all three measurement types into one test system while maintaining the measurement performance.

- IV measurements are a process to determine devices characteristics and performance. Depending on the device in hands we will attain different information.

- Lets consider a ohmic device, that is, is obeys Ohm law, V = RI then as the equation shows we would obtain a linear relation between voltage and current, as showed:

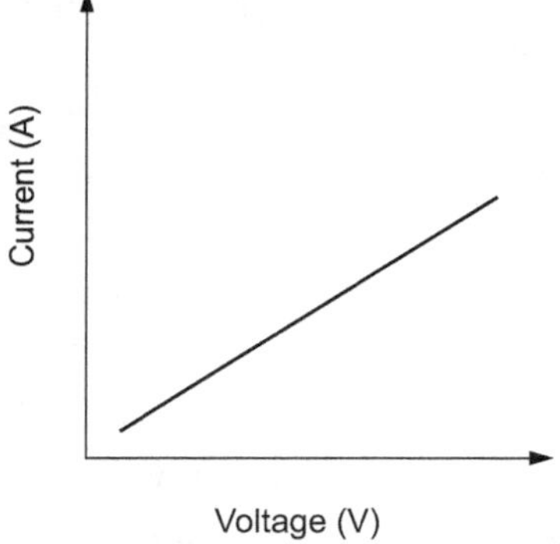



**Fig. 1.32 : Linear relation between voltage and current**

The resistance we measure has a contribution from the sample, the wires and the contacts:

$$R = R_{wires} + R_{contact} + R_{sample}$$

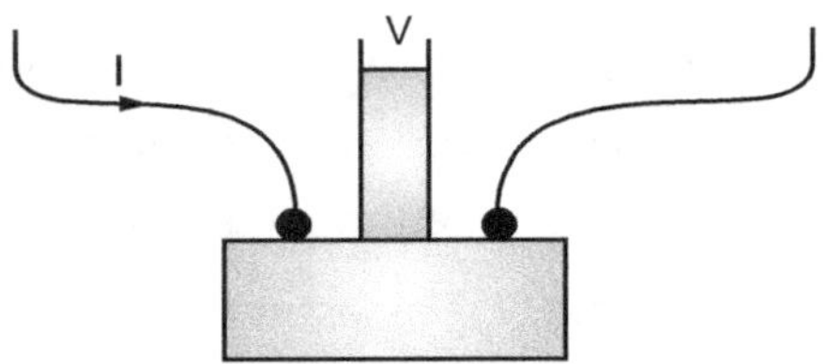



**Fig. 1.33 : Resistance measurement**

In order to minimise the contact and wires resistances the measurements are usually done with four probes; we are initially considering diodes. Below is sketch an IV curve. from observing the curve we can determine if we have a good quality device, since the bigger $\Delta$ is, better will the device perform. The threshold voltage is taken from the extrapolation to V=0 of the IV curve in forward bias.

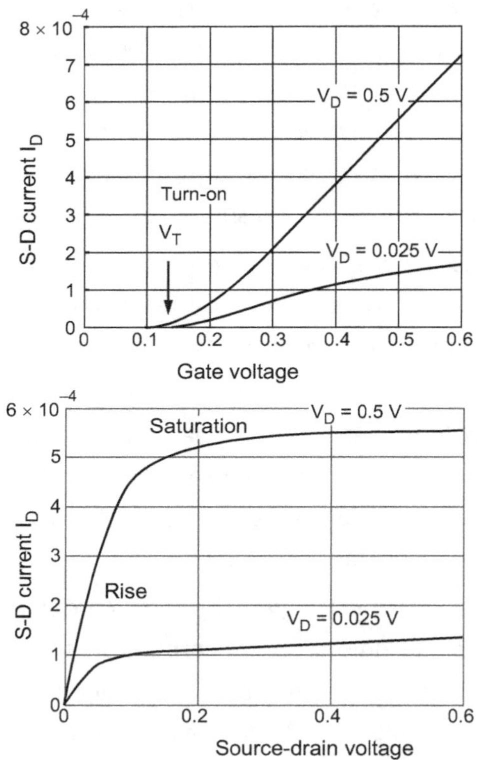



**Fig. 1.34: VI characteristics**

Another parameter that is possible to take is the Schoottky barrier height. A sketch of the band structure can be seen below:

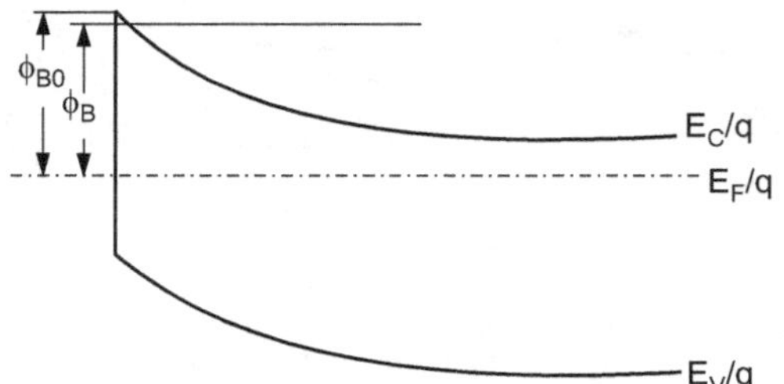



**Fig. 1.35 : Band structure of schoottkydiode**

In order to get the barrier height ($\varphi_B$) we need to consider the thermionic current voltage relationship:

$$I = I_s\left(\exp\frac{qV}{nkT} - 1\right)$$

where $I_s$ is the saturation current and

$$I_s = AA^*T^2\exp-\frac{q\phi B}{kT}$$

from this last equation we get

$$\phi_B = \frac{kT}{q}\ln\left(\frac{AA^*T^2}{I_s}\right)$$

with being the area of the diode and A*=120(m*/m). If we plot a semilog curve of $I_s$vs. V and extrapolate the curve to V = 0 we will get $\varphi_B$ for zero bias.

Lets consider now a MOSFET. Below is a sketch of one:

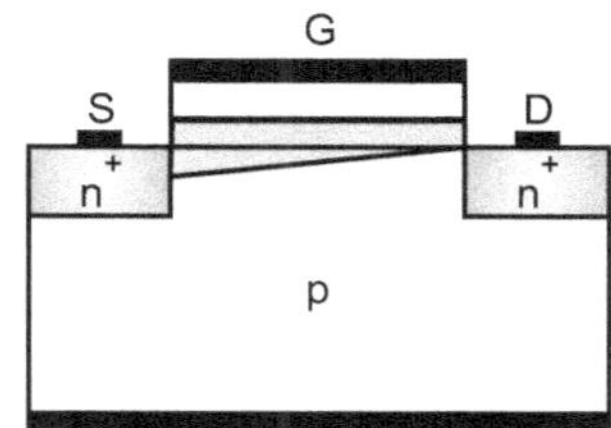


**Fig. 1.36: MOSFET**

where S is the source, D the drain and G the gate. One measure we can make is to measure the current at the drain and the source while we apply a constant voltage in the gate and a variable voltage along the drain and the source. What we will obtain is:

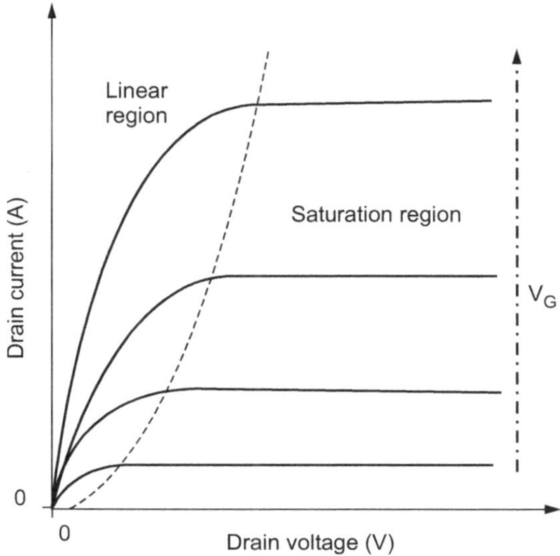


**Fig. 1.37 : Output Characteristics**

In above Fig. 1.37 shows we can see two different regions: the linear region and the saturation region. In the linear region the conduction channel acts like a resistor. if we continue to increase the voltage the channel potential will reduce the current, translating in the lost of linearity, if we continue to increase the voltage we will get to a point where the inversion charge reaches zero and we no longer see a change in the current, we reach the saturation regime.

The threshold voltage is the gate bias at which the devices turn on. And, it can be determine from the transfer characteristic, for that we need to measure the drain current while we change the gate voltage and extrapolate the curve to V=0. If we then plot the same curve in a semi log scale we will get the sub threshold swing.

There are many parameters could be extracted from IV measurement such as thersold voltage sub threshold slope and resistance which will discuss in next section.

**Capacitance–Voltage (C–V) Measurements**

For example, in order to measure effective field, it is common to use Split CV. With this technique we are able to determine charge depletion and charge inversion which are need.

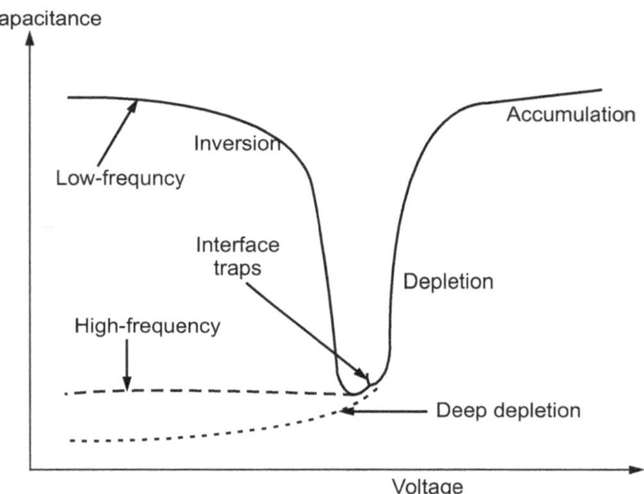


**Fig. 1.38: Capacitance Voltage characteristics**

In the gate channel configuration, $C_{gc,}$the substrate is connected to Earth while we apply voltage at the gate. The capacitance will be measured through the drain and source. The charge inversion will determine integrating the CV curve from the flatband voltage:

$$Q_{inv} = \frac{1}{WL}\int_{V_{FB}}^{V_G} C_{gc}d\,V_G$$

In the gate body configuration, $C_{gb,}$ the drain and the source are connected to the Earth while we apply voltage at the gate. the capacitance will now be measure through the substrate and the charge depletion will be determined integrating the CV curve, also, from the flatband voltage:

$$Q_{dep} = \frac{1}{WL}\int_{V_{FB}}^{V_G} C_{gc}d\,V_G$$

From the above expressions it can be perceive that another important parameter is the flatband voltage. This can be determined from the $1/C^2$ vs $V_G$ curve, by extrapolating the linear part to V = 0.

The effective field is given by

$$\varepsilon_{eff} = \frac{\eta\, Q_{int} + Q_{dep}}{k\varepsilon_0}$$

with $\eta$ = 1/2 for electron mobility and $\eta$ = 1/3 for hole mobility.

Combining this measurements with IV measurements we can see how mobility changes with the effective field since the mobility can be obtained from the drain conductance in the linear region:

$$\mu_{eff} = \frac{L}{W}\, \frac{gD}{q \cdot N_s}$$

where $q.N_s$ is the inversion carrier density determined through the gate channel capacitance.

- A variety of semiconductor parameters, capacitance measurements have been used on numerous different devices and structures. Three measurement techniques are used to derive critical parameters from a wide range of new materials, processes, devices. Multi–frequency capacitance offers capacitance vs. voltage (C–V), capacitance vs. frequency (C–f), and capacitance vs. time (C–t) measurements to evaluate at frequencies ranging from 10 MHz down to 1 kHz.

- Sometimes even lower frequency capacitance measurements are necessary to evaluate test parameters of thin film transistors, MEMS structures, and other high impedance devices is called very low frequency (VLF) C–V, this newer technique performs C–V measurements in the range of 10 mHz to 10 Hz.

- To characterize slow trapping and de–trapping phenomenon in some materials, a capacitance measurement technique called quasi static (or almost DC) measurements can be used.

## 1.5.2 Temperature Dependent Characterization

- Advanced technology regards the measurement of the temperature, Silicon (Si) power devices have been saturated in terms of higher temperature and higher power operation by advantage of their physical properties.

- Silicon Carbide (SiC) has been recognized as a material with the potential to replace Si devices due to their superior material advantages such as large band gap, high thermal conductivity, and high critical breakdown field strength.

- SiC devices are capable of operating at high voltages, high frequencies, and at higher junction temperatures. SiC unipolar devices such as Schottky diodes, VJFETs, MOSFETs, etc. have much higher breakdown voltages compared to their Si counterparts which makes them suitable for use in traction drives.

- The power devices in traction applications should be able to handle extreme environments which include a wide range of operating temperature.

**SiC Schottky Diodes**

- The Schottky diode (named after German physicist Walter H. Schottky), also known as hot carrier diode, is a semiconductor diode with a low forward voltage drop and a very fast switching action. The cat's–whisker detectors used in the early days of wireless and metal rectifiers used in early power applications can be considered primitive Schottky diodes.

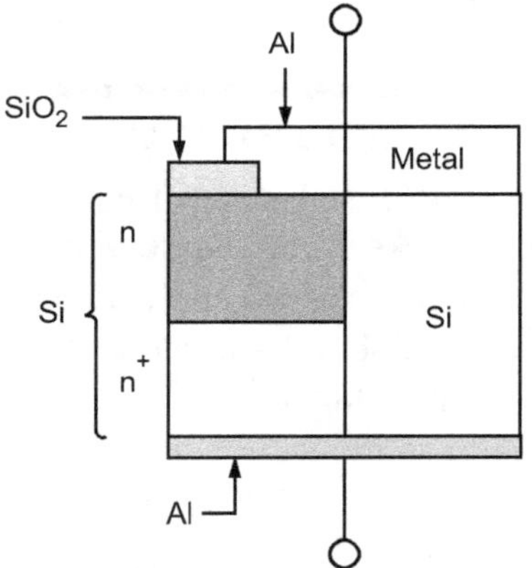


**Fig. 1.39 : Construction of Schottky Diode**

**Fig. 1.40 : Symbol of Schottky diode**

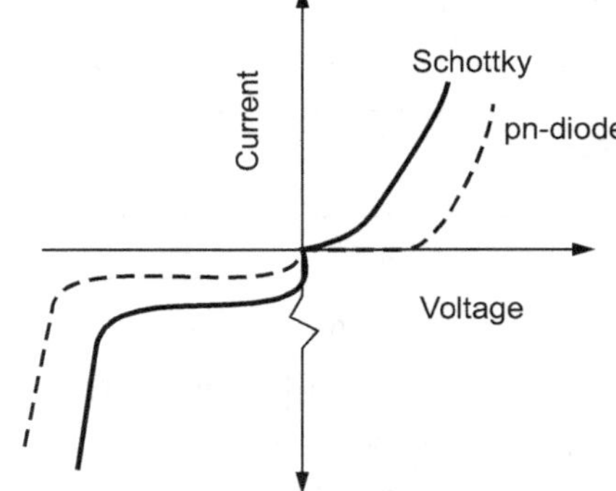


**Fig. 1.41 : VI characteristics of Schottky diode**

- When forward current flows through a solid–state diode, there is a small voltage drop across its terminals.

A silicon diode has a typical voltage drop of 0.6–0.7 V, while a Schottky diode has a voltage drop of 0.15–0.45 V. This lower voltage drop can be used to give higher switching speeds and better system efficiency.

- A metal–semiconductor junction is formed between a metal and a semiconductor, creating a Schottky barrier (instead of asemiconductor–semiconductor junction as in conventional diodes). Typical metals used are molybdenum, platinum, chromium or tungsten, and certain silicides (e.g., palladium silicide and platinum silicide), whereas the semiconductor would typically be n–type silicon. The metal side acts as the anode, and n–type semiconductor acts as the cathode of the diode. This Schottky barrier results in both very fast switching and low forward voltage drop.

- The combination of the metal and semiconductor determines the forward voltage of the diode. Both n–type and p–type semiconductors can develop Schottky barriers. However, the p–type typically has a much lower forward voltage. As the reverse leakage current increases dramatically with lowering the forward voltage, it can not be too low, so the usually employed range is about 0.5–0.7 V, and p–type semiconductors are employed only rarely. Titanium silicide and other refractory silicides, which are able to withstand the temperatures needed for source/drain annealing in CMOS processes, usually have too low a forward voltage to be useful, so processes using these silicides therefore usually do not offer Schottky diodes.

- With increased doping of the semiconductor, the width of the depletion region drops. Below a certain width, the charge carriers can tunnel through the depletion region. At very high doping levels, the junction does not behave as a rectifier anymore and becomes an ohmic contact. This can be used for the simultaneous formation of ohmic contacts and diodes, as a diodes will form between the silicide and lightly doped n–type region, and an ohmic contact will form between the silicide and the heavily doped n– or p–type region. Lightly doped p–type regions pose a problem, as the resulting contact has too high a resistance for a good ohmic lapis contact, but too low a forward voltage and too high a reverse leakage to make a good diode.

- As the edges of the Schottky contact are fairly sharp, a high electric field gradient occurs around them, which limits how large the reverse breakdown voltage threshold can be. Various strategies are used, from guard rings to overlaps of metallization to spread out the field gradient. The guard rings consume valuable die area and are used primarily for larger higher–voltage diodes, while overlapping metallization is employed primarily with smaller low–voltage diodes.

- Schottky diodes are often used as antisaturation clamps in Schottky transistors. Schottky diodes made from palladium silicide (PtSi) are excellent due to their lower forward voltage (which has to be lower than the forward voltage of the base–collector junction). The Schottky temperature coefficient is lower than the coefficient of the B–C junction, which limits the use of PtSi at higher temperatures.

- For power Schottky diodes, the parasitic resistances of the buried n+ layer and the epitaxial n–type layer become important. The resistance of the epitaxial layer is more important than it is for a transistor, as the current must cross its entire thickness. However, it serves as a distributed ballasting resistor over the entire area of the junction and, under usual conditions, prevents localized thermal runaway.

- In comparison with the power p–n diodes the Schottky diodes are less rugged. The junction is direct contact with the thermally sensitive metallization, a Schottky diode can therefore dissipate less power than an equivalent–size, p–n diodes counterpart with a deep–buried junction before failing (especially during reverse breakdown). The relative advantage of the lower forward voltage of Schottky diodes is diminished at higher forward currents, where the voltage drop is dominated by the series resistance.

**Key Specification Parameters**

In view of the particular properties of the Schottky diode there are several parameters that are of key importance when determining the operation of one of these diodes against the more normal PN junction diodes.

- **Forward Voltage Drop:** In view of the low forward voltage drop across the diode, this is a parameter that is of particular concern. As can be seen from the Schottky diode IV characteristic, the voltage across the diode varies according to the current being carried. Accordingly any specification given provides the forward voltage drop for a given current. Typically the turn–on voltage is assumed to be around 0.2 V.

- **Reverse Breakdown:** Schottky diodes do not have a high breakdown voltage. Figures relating to this include the maximum Peak Reverse Voltage, maximum Blocking DC Voltage and other similar parameter names. If these figures are exceeded then there is a possibility the diode will enter reverse breakdown. It should be noted that the RMS value for any voltage will be $\frac{1}{\sqrt{2}}$ times the constant value. The upper limit for reverse breakdown is not high when compared to normal PN junction diodes. Maximum figures, even for rectifier diodes only reach around 100 V. Schottky diode rectifiers seldom exceed this value because devices that would operate above this value even by moderate amounts would exhibit forward voltages equal to or greater than equivalent PN junction rectifiers.

- **Capacitance:** The capacitance parameter is one of great importance for small signal RF applications. Normally the junctions areas of Schottky diodes are small and therefore the capacitance is small. Typical values of a few pico Farads are normal. As the capacitance is dependent upon any depletion areas, etc, the capacitance must be specified at a given voltage.

- **Reverse Recovery Time:** This parameter is important when a diode is used in a switching application. It is the time taken to switch the diode from its forward conducting or 'ON' state to the reverse 'OFF' state. The charge that flows within this time is referred to as the reverse recovery charge. The time for this parameter for a Schottky diode is normally measured in nanoseconds, ns. Some exhibit times of 100 ps. In fact what little recovery time is required mainly arises from the capacitance rather than the majority carrier recombination. As a result there is very little reverse current overshoot when switching from the forward conducting state to the reverse blocking state.

- **Working Temperature:** The maximum working temperature of the junction, $T_j$ is normally limited to between 125 to 175°C. This is less than that which can be sued with ordinary silicon diodes. Care should be taken to ensure heat sinking of power diodes does not allow this figure to be exceeded.

- **Reverse Leakage Current:** The reverse leakage parameter can be an issue with Schottky diodes. It is found that increasing temperature significantly increases the reverse leakage current parameter. Typically for every 25°C increase in the diode junction temperature there is an increase in reverse current of an order of magnitude for the same level of reverse bias.

**Limitations:**

- The most evident limitations of Schottky diodes are their relatively low reverse voltage ratings, and their relatively high reverse leakage current. For silicon–metal Schottky diodes, the reverse voltage is typically 50 V or less. Some higher–voltage designs are available (200 V is considered a high reverse voltage). Reverse leakage current, since it increases with temperature, leads to a thermal instability issue. This often limits the useful reverse voltage to well below the actual rating.

- While higher reverse voltages are achievable, they would present a higher forward breakdown voltage, comparable to other types of standard diodes. Such Schottky diodes would have no advantage unless great switching speed is required.

**Silicon Carbide Schottky Diode**

- Schottky diodes constructed from silicon carbide have a much lower reverse leakage current than silicon Schottky diodes, as well as higher forward voltage and reverse voltage. As of 2011 they were available from manufacturers in variants up to 1700 V of reverse voltage.

- Silicon carbide has a high thermal conductivity, and temperature has little influence on its switching and thermal characteristics. With special packaging, silicon carbide Schottky diodes can operate at junction temperatures of over 500 K (about 200 °C), which allows passive radiative cooling in aerospace applications.

**Applications**

1. **Voltage Clamping**

- While standard silicon diodes have a forward voltage drop of about 0.6 V and germanium diodes 0.2 V, Schottky diodes' voltage drop at forward biases of around 1mA is in the range of 0.15 V to 0.46 V, which makes them useful in voltage clamping applications and prevention of transistor saturation. Due to the higher current density in the Schottky diode.

2. **Reverse Current and Discharge Protection**

- A Schottky diode's low forward voltage drop, less energy is wasted as heat making them the most efficient choice for applications sensitive to efficiency. For instance, they are used in stand–alone ("off–grid") photovoltaic (PV) systems to prevent batteries from discharging through the solar panels at night, called "blocking diodes".

- They are also used in grid–connected systems with multiple strings connected in parallel, in order to prevent reverse current flowing from adjacent strings through shaded strings if the "bypass diodes" have failed.

## Switched–Mode Power Supplies

- Schottky diode are also used as rectifiers in switched–mode power supplies. The low forward voltage and fast recovery time leads to increased efficiency.

- They can also be used in power supply "OR"ing circuits in products that have both an internal battery and a mains adapter input, or similar. However, the high reverse leakage current presents a problem, in this case, as any high–impedance voltage sensing circuit (e.g., monitoring the battery voltage or detecting whether a mains adapter is present) will see the voltage from the other power source through the diode leakage.

- SiC Schottky diodes are majority carrier devices and are attractive for high frequency applications because they have lower switching losses compared to p–n diodes. However, Schottky diodes have higher leakage currents, which affect the breakdown voltage rating of the device.

## Characteristics of Schottky Diode

- **Temperature affects every Semiconductor Device**, and diodes are no exception. Temperature can have considerable effect on the characteristics of diode. The goal of this section is to understand how temperature affects the characteristics of diode. We shall study the effects of temperature on both forward and reverse characteristics.

- We studied about the characteristics of diode by applying Shockley's equation. Shockley's equation is as follows.

$$I_D = I_S \left( e^{V_D/\eta V_J} - 1 \right)$$

- Consider the term $V_T$ given in above equation. This term is called thermal voltage and is dependent on temperature by the relation $V_T = kT/q$, where k is Boltzmann's constant, q is the charge on an electron and T is the temperature in Kelvin. **This term**

**indirectly suggests the dependence of diode characteristics on temperature.**

**Effect of Temperature on Forward Characteristics :**

- The characteristics curve of a Si diode shifts to the left at the rate of –2.5 mV per degree centigrade change in temperature in forward bias region.

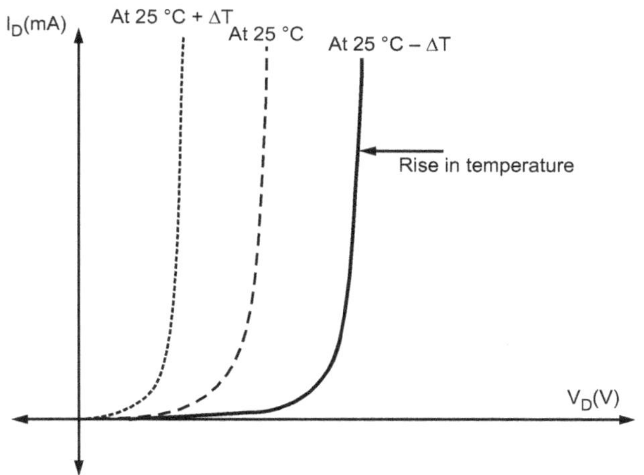


**Fig. 1.42 : Effect of temperature on forward bias**

- Above Fig. 1.42 shows the curves are shown far apart just for illustration purpose and are not to scale. The curve shifts to the left at the rate of –2.5 mV per degree centigrade change in temperature. Hence if the temperature increases from room temperature (25° C) to 80° C, the voltage drop across the diode will be $(80 - 25) \times 2.5$ mV = 137.5 mV.

**Effect of Temperature on Reverse Characteristics :**

- In the reverse bias region, the reverse saturation current of Si and Ge diodes doubles for every 10° C rise in temperature. Let us take an example to understand how much the reverse saturation current changes with temperature. Consider an increase of temperature from 25°C to 85°C, where the reverse saturation current at 25°C is 100 nA. The temperature increases by 60°C (25°C to 85°C), which is $6 \times 10$. Hence the reverse saturation current would increase by a factor of 26 = 64. Hence the reverse saturation current at 85 °C will be 100 nA $\times$ 64 = 6400 nA.

- A showing in Fig. 1.43 the variation of reverse saturation current with temperature is shown below. The difference between the curve is exaggerated for illustration purpose.

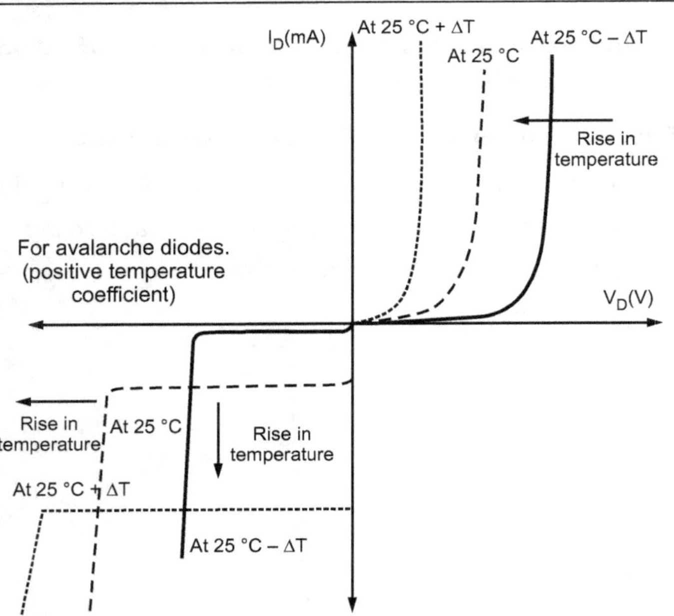



**Fig. 1.43: Effect of temperature on Reverse bias**

- From the above Fig. 1.43 shows that the reverse saturation current increases with increase in temperature. The graph also shows how the reverse breakdown voltage changes with temperature. It is indicated in the above graph that the reverse breakdown voltage increases with an increase in temperature. However, it is only true for avalanche diodes. The reverse breakdown voltage for Zener diodes decreases with an increase in temperature, which is shown in the Fig. 1.44 below.

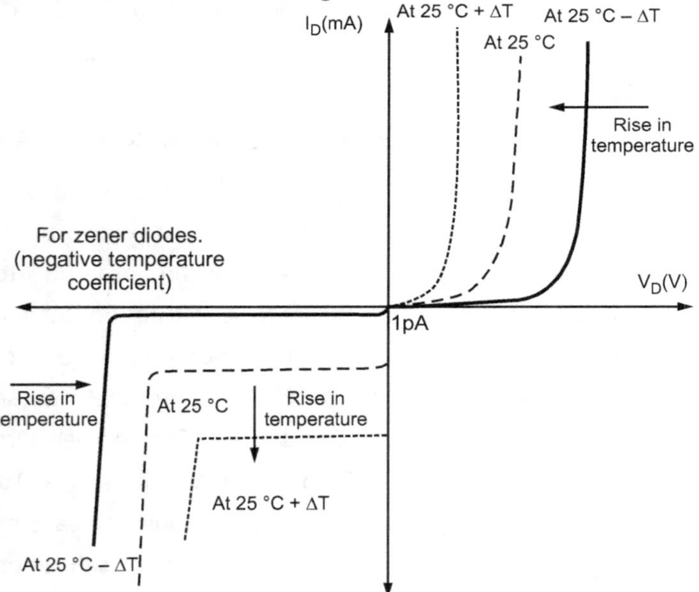



**Fig 1.44: Effect of temperature on Zener diode**

- The threshold voltage or the knee voltage are the on–state resistance is different for the diodes because of differences in device dimensions for different ratings. The static characteristics of diodes for a temperature range of –50 °C to 175 °C. The on–state voltage drop

of a Schottky diode is dependent on barrier height and the on–state resistance. Both parameters vary with temperature and hence contribute to the temperature dependence of forward characteristics.

- At lower current levels the built–in potential (barrier potential) decreases with increasing temperature due to reduction in barrier height.

- As the temperature increases, the thermal energy of electrons increase which causes lowering of the barrier height.

**Dynamic Characteristics**

- A buck chopper with an inductive load is built for evaluating the switching characteristics of the diodes. An IGBT is used as the main switch and is switched at 20 kHz with a 25% duty ratio.

- The power losses for various forward peak currents and different temperatures are shown for the Si diode and diode $S_4$. The power losses for the Si diode increase with temperature, because of the increase in peak reverse recovery current.

- The switching loss for diode $S_4$ is almost independent of the change in temperature. The reverse recovery current is dependent on charge stored in the drift region.

- The SiC Schottky diode has no stored charge because it is a majority carrier device, and hence has virtually constant turn on energy loss for a wide temperature range.

- The reverse recovery current reduces the oscillation due to ringing and sunbber circuit for the limiting recovery. As a result of this condition efficiency as the losses or minimized.

- The blocking layer of thickness, due to the high electric breakdown field of the SiC material, contributes to the low switching losses of the SiC diode.

**SiC Vertical JFET (VJFET)**

- We saw previously that a bipolar junction transistor is constructed using two PN–junctions in the main current carrying path between the emitter and the collector terminals. The **Junction Field Effect Transistor** (JUGFET or JFET) has no PN–junctions but instead has a narrow piece of high resistivity semiconductor material forming a "channel" of either N–type or P–type silicon for the majority carriers to flow through with two ohmic electrical connections at either end commonly called the **drain** and the **source** respectively.

- There are two basic configurations of junction field effect transistor, the N–channel JFET and the P–channel JFET. The N–channel JFET's channel is doped with donor impurities meaning that the flow of current through the channel is negative (hence the term N–channel) in the form of electrons.

- Likewise, the P–channel JFET's channel is doped with acceptor impurities meaning that the flow of current through the channel is positive (hence the term P–channel) in the form of holes. N–channel JFET's have a greater channel conductivity (lower resistance) than their equivalent P–channel types, since electrons have a higher mobility through a conductor compared to holes. This makes the N–channel JFET's a more efficient conductor compared to their P–channel counterparts.

- We have said previously that there are two ohmic electrical connections at either end of the channel called the **drain** and the **source**. But within this channel there is a third electrical connection which is called the **gate** terminal and this can also be a P–type or N–type material forming a PN–junction with the main channel. The relationship between the connections of a junction field effect transistor and a bipolar junction transistor are compared below.

### Table 1.2 : Comparison of Connections between a JFET and a BJT

| Bipolar Transistor | Field Effect Transistor |
|---|---|
| Emitter – (E) >> Source – (S) | |
| Base – (B) >> Gate – (G) | |
| Collector – (C) >> Drain – (D) | |

The symbols and basic construction for both configurations of JFETs are shown in Fig. 1.45 below.

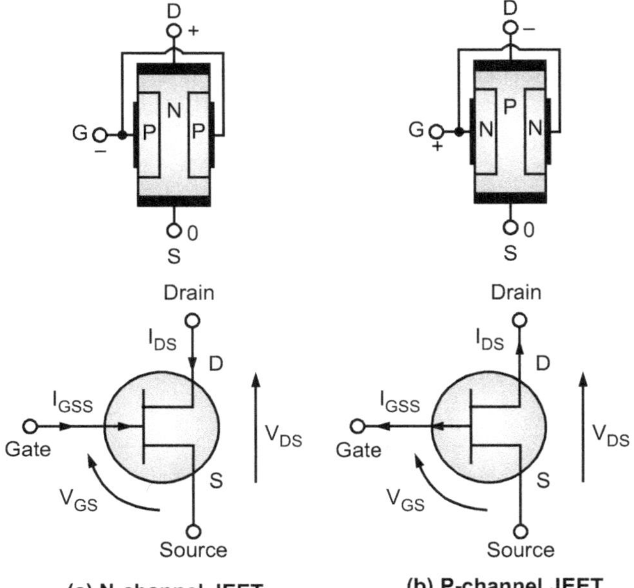

(a) N-channel JEET      (b) P-channel JEET

**Fig 1.45: Construction of JFET**

- The semiconductor "channel" of the **Junction Field Effect Transistor** is a resistive path through which a voltage $V_{DS}$ causes a current $I_D$ to flow and as such the junction field effect transistor can conduct current equally well in either direction. As the channel is resistive in nature, a voltage gradient is thus formed down the length of the channel with this voltage becoming less positive as we go from the drain terminal to the source terminal.

- The result is that the PN–junction therefore has a high reverse bias at the drain terminal and a lower reverse bias at the source terminal. This bias causes a "depletion layer" to be formed within the channel and whose width increases with the bias.

- The magnitude of the current flowing through the channel between the drain and the Source terminals is controlled by a voltage applied to the gate terminal, which is a reverse–biased. In an N–channel JFET this Gate voltage is negative while for a P–channel JFET the Gate voltage is positive. The main difference between the JFET and a BJT device is that when the JFET junction is reverse–biased the Gate current is practically zero, whereas the Base current of the BJT is always some value greater than zero.

- The above Fig. 1.46 shows an N–type semiconductor channel with a P–type region called the gate diffused into the N–type channel forming a reverse biased PN–junction and it is this junction which forms the *depletion region* around the gate area when no external voltages are applied. JFETs are therefore known as depletion mode devices.

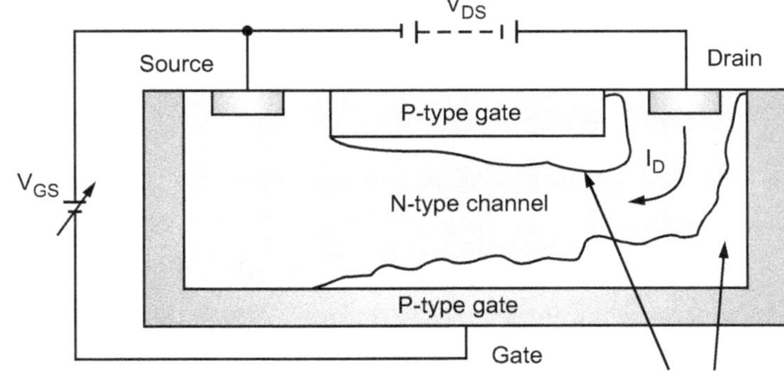

**Fig 1.46: Biasing on N–channel JFET**

- This depletion region produces a potential gradient which is of varying thickness around the PN–junction and restrict the current flow through the channel by reducing its effective width and thus increasing the overall resistance of the channel itself.

- Then we can see that the most–depleted portion of the depletion region is in between the gate and the drain, while the least–depleted area is between the gate and the source. Then the JFET's channel conducts with zero bias voltage applied (ie, the depletion region has near zero width).

- With no external Gate voltage ($V_G = 0$), and a small voltage ($V_{DS}$) applied between the drain and the Source, maximum saturation current ($I_{DSS}$) will flow through the channel from the drain to the Source restricted only by the small depletion region around the junctions.

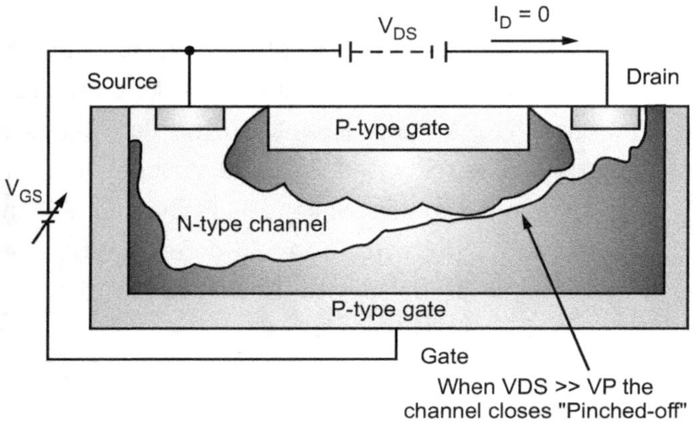


**Fig 1.47: JFET Channel Pinched–off**

- If a small negative voltage ($-V_{GS}$) is now applied to the gate the size of the depletion region begins to increase reducing the overall effective area of the channel and thus reducing the current flowing through it, a sort of "squeezing" effect takes place. So by applying a reverse bias voltage increases the width of the depletion region which in turn reduces the conduction of the channel.

- Since the PN–junction is reverse biased, little current will flow into the gate connection. As the Gate voltage ($-V_{GS}$) is made more negative, the width of the channel decreases until no more current flows between the drain and the source and the FET is said to be "pinched–off" (similar to the cut–off region for a BJT). The voltage at which the channel closes is called the "pinch–off voltage", ($V_P$).

- In this pinch–off region the gate voltage, $V_{GS}$ controls the channel current and $V_{DS}$ has little or no effect.

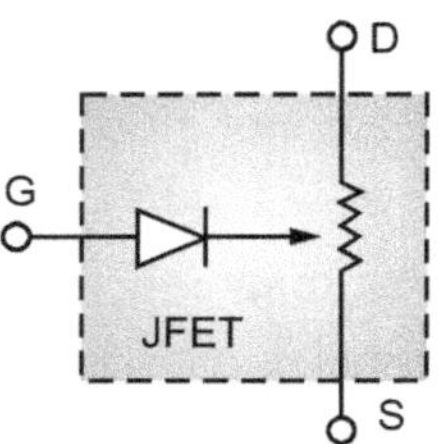


**Fig. 1.48: JFET Model**

- The result is that the FET acts more like a voltage controlled resistor which has zero resistance when $V_{GS} = 0$ and maximum "ON" resistance ($R_{DS}$) when the gate voltage is very negative. Under normal operating conditions, the JFET gate is always negatively biased relative to the source.

- It is essential that the gate voltage is never positive since if it is all the channel current will flow to the gate and not to the Source, the result is damage to the JFET. Then to close the channel:

  1. No Gate voltage ( $V_{GS}$ ) and $V_{DS}$ is increased from zero.

  2. No $V_{DS}$ and Gate control is decreased negatively from zero.

  3. $V_{DS}$ and $V_{GS}$ varying.

- The P–channel **Junction Field Effect Transistor** operates the same as the N–channel above, with the following exceptions:

  1. Channel current is positive due to holes,

  2. The polarity of the biasing voltage needs to be reversed.

- The output characteristics of an N–channel JFET with the gate short–circuited to the source is as shown in following Fig. 1.49.

**Output characteristic V–I curves of a typical junction FET.**

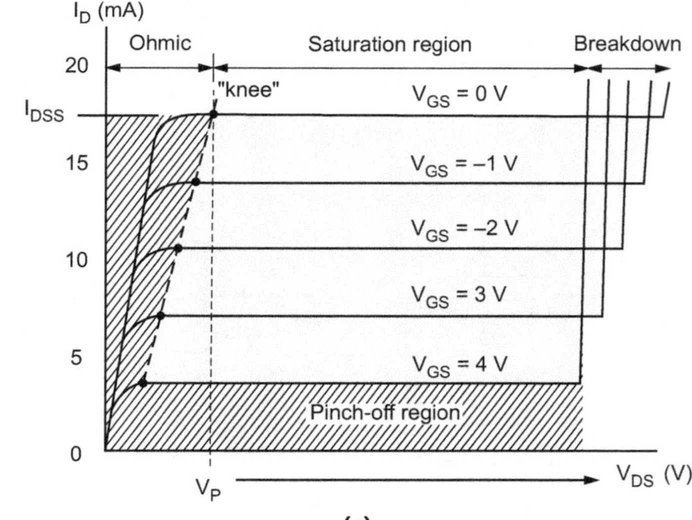

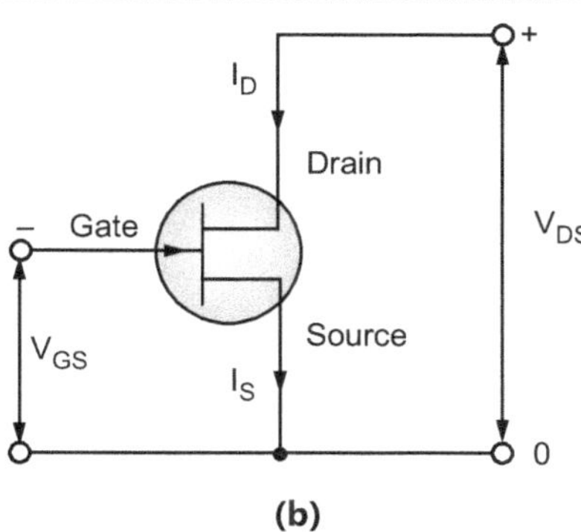



**(b)**

**Fig 1.49: Output Characteristics of JFET**

The voltage $V_{GS}$ applied to the gate controls the current flowing between the drain and the source terminals. $V_{GS}$ refers to the voltage applied between the gate and the source while $V_{DS}$ refers to the voltage applied between the drain and the Source.

Because a **Junction Field Effect Transistor** is a voltage controlled device, "No current flows into the gate!" then the source current ( $I_S$ ) flowing out of the device equals the drain current flowing into it and therefore ( $I_D = I_S$ ).

The characteristics curves example shown in above in Fig. 1.49 (a) shows the four different regions of operation for a JFET and these are given as:

1. **Ohmic Region :** When $V_{GS} = 0$ the depletion layer of the channel is very small and the JFET acts like a voltage controlled resistor.

2. **Cut–off Region :** This is also known as the pinch–off region were the gate voltage, $V_{GS}$ is sufficient to cause the JFET to act as an open circuit as the channel resistance is at maximum.

3. **Saturation or Active Region :** The JFET becomes a good conductor and is controlled by the Gate–Source voltage, ( $V_{GS}$ ) while the drain–source voltage, ( $V_{DS}$ ) has little or no effect.

4. **Breakdown Region :** The voltage between the drain and the source, ( $V_{DS}$ ) is high enough to causes the JFET's resistive channel to break down and pass uncontrolled maximum current.

The characteristics curves for a P–channel junction field effect transistor are the same as those above, except that the drain current $I_D$ decreases with an increasing positive gate–source voltage, $V_{GS}$.

The drain current is zero when $V_{GS} = V_P$. For normal operation, $V_{GS}$ is biased to be somewhere between $V_P$ and 0. Then we can calculate the drain current, $I_D$ for any given bias point in the saturation or active region as follows:

Drain current in the active region.

$$I_D = I_{DSS}\left[1 - \frac{V_{GS}}{V_P}\right]^2$$

Note that the value of the Drain current will be between zero (pinch–off) and $I_{DSS}$ (maximum current). By knowing the Drain current $I_D$ and the drain–source voltage $V_{DS}$ the resistance of the channel ($I_D$) is given as:

Drain–Source channel resistance.

$$R_{DS} = \frac{\Delta V_{DS}}{\Delta I_D} = \frac{1}{g_m}$$

Where $g_m$ is the "transconductance gain" since the JFET is a voltage controlled device and which represents the rate of change of the Drain current with respect to the change in Gate–Source voltage.

**Modes of FET's**

- Like the bipolar junction transistor, the field effect transistor being a three terminal device is capable of three distinct modes of operation and can therefore be connected within a circuit in one of the following configurations.

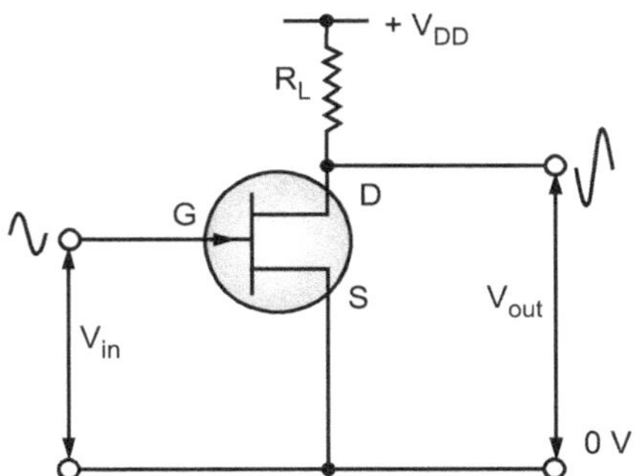



**Fig 1.50: Common Source (CS) Configuration**

- In the **Common Source** configuration (similar to common emitter), the input is applied to the gate and its output is taken from the drain as shown. This is the most common mode of operation of the FET due to its high input impedance and good voltage amplification and as such common source amplifiers are widely used.

- The common source mode of FET connection is generally used audio frequency amplifiers and in high input impedance pre–amps and stages. Being an amplifying circuit, the output signal is 180° "out–of–phase" with the input.

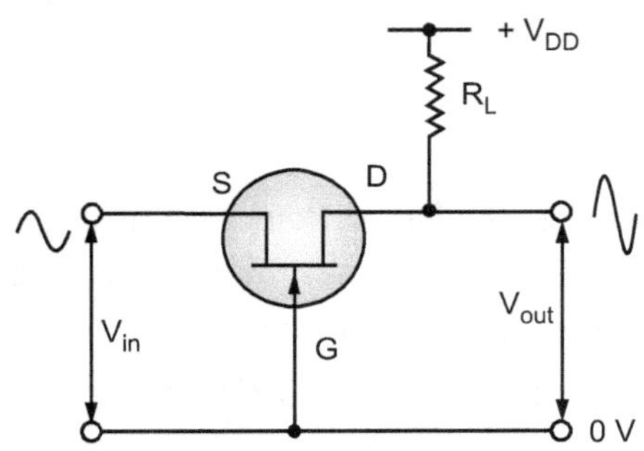

**Fig. 1.51: Common Gate (CG) Configuration**

- In the **Common Gate** configuration (similar to common base), the input is applied to the Source and its output is taken from the drain with the gate connected directly to ground (0V) as shown. The high input impedance feature of the previous connection is lost in this configuration as the common gate has a low input impedance, but a high output impedance.

- This type of FET configuration can be used in high frequency circuits or in impedance matching circuits were a low input impedance needs to be matched to a high output impedance. The output is "in–phase" with the input.

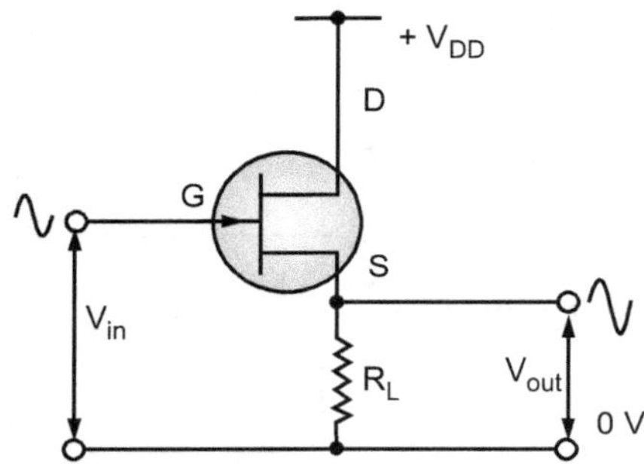

**Fig 1.52: Common Drain (CD) Configuration**

- In the **Common Drain** configuration (similar to common collector), the input is applied to the gate and its output is taken from the source. The common drain or "source follower" configuration has a high input impedance and a low output impedance and near–unity voltage gain so is therefore used in buffer amplifiers. The voltage gain of the source follower configuration is less than unity, and the output signal is "in–phase", $0°$ with the input signal.

- This type of configuration is referred to as "Common Drain" because there is no signal available at the drain connection, the voltage present, $+V_{DD}$ just provides a bias. The output is in–phase with the input.

## The JFET Amplifier

- Just like the bipolar junction transistor, JFET's can be used to make single stage class A amplifier circuits with the JFET common source amplifier and

characteristics being very similar to the BJT common emitter circuit. The main advantage JFET amplifiers have over BJT amplifiers is their high input impedance which is controlled by the Gate biasing resistive network formed by $R_1$ and $R_2$ as shown.

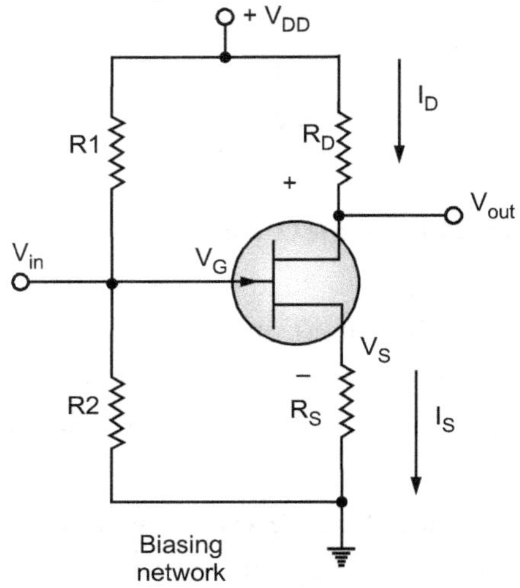

**Fig 1.53: JFET Amplifier**

$$V_S = I_D R_S = \frac{V_{DD}}{4}$$

$$V_S = V_G - V_{GS}$$

$$V_G = \left(\frac{R_2}{R_1 + R_2}\right) V_{DD}$$

$$ID = \frac{V_S}{R_S} = \frac{V_G - V_{GS}}{R_S}$$

- This Common Source (CS) amplifier circuit is biased in class "A" mode by the voltage divider network formed by resistors $R_1$ and $R_2$. The voltage across the source resistor $R_S$ is generally set to be about one quarter of $V_{DD}$, ($V_{DD}$ /4). The required gate voltage can then be calculated using this $R_S$ value. Since the gate current is zero, ($I_G = 0$) we can set the required DC quiescent voltage by the proper selection of resistors $R_1$ and $R_2$.

- The control of the drain current by a negative gate potential makes the Junction Field Effect Transistor useful as a switch and it is essential that the gate voltage is never positive for an N–channel JFET as the channel current will flow to the gate and not the drain resulting in damage to the JFET. The principals of operation for a P–channel JFET are the same as for the N–channel JFET, except that the polarity of the voltages need to be reversed.

- Transistors, we will look at another type of Field Effect Transistor called a MOSFET whose gate connection is completely isolated from the main current carrying channel.

- JFET, a unipolar device, has several advantages compared to MOSFET devices. It has low voltage drop and higher switching speed. It is free from the gate oxide interface problems not like the MOSFET.

- JFET is a normally–on device and conducts even though there is no gate voltage applied. Therefore it requires protection circuit for system power failures to prevent a short circuit.

- This normally–on feature demands special gate drive designs increasing the complexity of design. The VJFET can be used in high current and voltage applications, unlike Si JFET because of the vertical structure and the intrinsic properties of SiC.

- A normally–on SiC VJFET rated at 1200V and 2A was tested to study the high temperature behavior of the device. The on–resistance of the VJFET increases from 0.36 at – 50 °C to 1.4 at 175 °C.

- The values of the on–resistance are high; however, these devices have positive temperature coefficient which enables easy paralleling of devices and the on–state resistance would decrease

**Transfer Characteristics**

**JFET Output Characteristic**

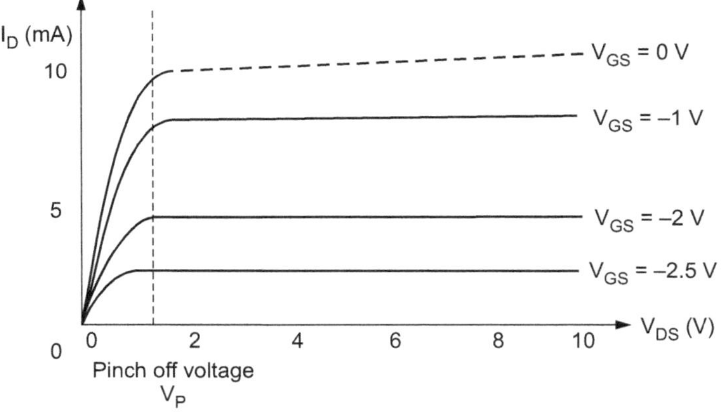


**Fig 1.54: JFET Output Characteristics**

- In the JFET output characteristics shown in Fig. 1.54, the Drain current $I_D$ shows very little change, and the curves are very nearly horizontal at voltages greater than the pinch off voltage. Almost all of the expected increase in current, due to the increase in voltage between Source and Drain ($V_{DS}$), is offset by the narrowing of the conducting channel due to the growing depletion layers.

- The transfer characteristic for a JFET, which shows the change in drain current ($I_D$) for a given change in gate–source voltage ($V_{GS}$), is shown in Fig 1.55. Because the JFET input (the gate) is voltage operated, the gain of the transistor cannot be called current gain,

as with bipolar transistors. The drain current is controlled by the gate–source voltage, so the graph shows milliamperes per volt (mA / V), and as I / V is conductance (the inverse of resistance V / I) the slope of this graph (the gain of the device) is called the forward or mutual transconductance, which has the symbol $g_m$. Therefore the higher the value of $g_m$ the greater the amplification.

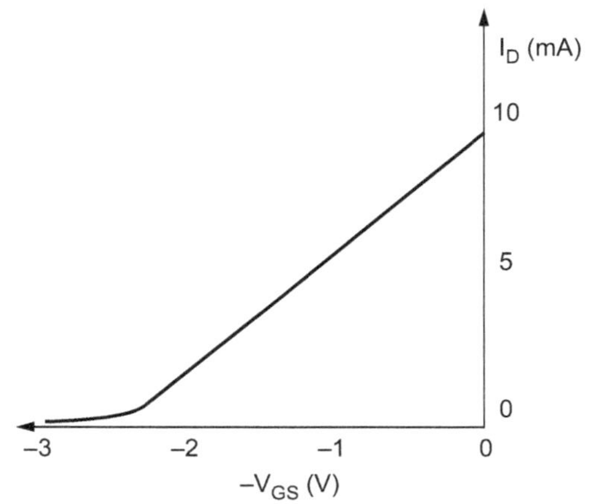


**Fig. 1.55 : JFET Transfer Characteristic**

- Notice that $V_{GS}$ is always shown as being negative; in reality it may be zero or slightly above zero, but the gate is always more negative than the N-type channel between source and drain. Note also that the slope of the curve in the transfer characteristic is less steep than that of the transfer characteristic for a typical bipolar transistor. This means that a JFET will have a lower gain than that of a bipolar transistor.

- This disadvantage is offset by the advantage of having an extremely high input resistance. A typical input resistance for a JFET would be in the region of $1 \times 10^{10}$ ohms (10,000 Megohms!) compared with 2K to 3K Ohms for a bipolar device.

- This makes the JFET ideal for applications where the circuit or device driving the JFET amplifier cannot supply any appreciable current, an example being the Electret microphone, which uses a FET within the microphone to amplify the tiny voltage variations appearing across the vibrating diaphragm element.

- Another feature of the JFET that makes it more suited to very high frequency use than bipolar transistors, is the absence of junctions in the JFET conducting channel. In a bipolar transistor two PN junctions forming tiny capacitances, exist between base and emitter, and base and collector, due to the PN junctions. These capacitances will limit high frequency performance, as they provide negative feedback paths at high frequencies. Because the JFET is in effect just a slab of silicon between Source and Drain, the stray capacitances that exist in bipolar devices are absent, so high frequency performance is improved, making JFETs usable even at hundreds of MHz.

- The negative gate pinch–off voltage required to turn–off the device does not vary much with an increase in drain to source voltage $V_{ds}$.

- This pinch–off voltage determines the voltage requirement of the gate drive circuit. A gate drive was developed for the VJFET for high frequency and high temperature operation. Based on the transfer characteristics, the gate drive was designed for a voltage of 30V and a 250 kHz operation was achieved for peak gate currents of 0.8 A.

**Gate Drive Requirements**

- SiC VJFET switches can be operated at higher switching frequencies and higher temperatures; therefore, they have different gate drive requirements than traditional Si power switches.

- SiC VJFET are devices can be turned OFF by applying a negative voltage this condition requires the higher than the typical switch Si.

- One of the important parameters in gate drive design for VJFETs is that they have a large stray capacitance between the gate and the other terminals.

- Total input capacitance of VJFET, CISS, determines the current required by the gate and the rate at which the applied gate voltage is built across the gate and source terminals.

- These capacitances are due to the geometrical design of the device and also because of the parasitic body diode.

- Therefore, the circuit that drives the gate terminal should be capable of supplying a reasonable current so the stray capacitance can be charged up as quickly as possible.

- The gate drive was tested with several capacitors as load before testing the driver circuit with the device.

- When these are operated at high frequencies, they also need high peak gate currents. The CISS of VJFET is higher and a resistor is added between the gate driver and the source terminal.

- This is to limit the current to the gate so that the maximum voltage on the gate terminal does not overshoot.



- The characterization of nanoparticles is a branch of nanometrology that deals with the characterization of physical and chemical properties of nanoparticles. Nanoparticles have at least one primary external dimension of less than 100 nanometers, and are often engineered for their unique properties.

- Nanoparticles are unlike conventional chemicals in that their chemical composition and concentration are not sufficient metrics for a complete description, because they vary in other physical properties such as size, shape, surface properties, crystallinity, and dispersion state.

- Nanoparticles are characterized for various purposes, including nanotoxicology studies and exposure assessment in workplaces to assess their health and safety hazards, as well as manufacturing process control.

- There is a wide range of instrumentation to measure these properties, including microscopy and spectroscopy methods as well as particle counters. Metrology standards and reference materials for nanotechnology, while still a new discipline, are available from many organizations.

### 1.6.1 Types of Nano Characterization

- Microscopy methods generate images of individual nanoparticles to characterize their shape, size, and location. Electron microscopy and scanning probe microscopy are the dominant methods. Because nanoparticles have a size below the diffraction limit of visible light, conventional optical microscopy is not useful. Electron microscopes can be coupled to spectroscopic methods that can perform elemental analysis.

- Microscopy methods are destructive, and can be prone to undesirable artifacts from sample preparation such as drying or vacuum conditions required for some methods, or from probe tip geometry in the case of scanning probe microscopy.

- Additionally, microscopy is based on single-particle measurements, meaning that large numbers of individual particles must be characterized to estimate their bulk properties. A newer method, enhanced dark-field microscopy with hyperspectral imaging, shows promise for imaging nanoparticles in complex matrices such as biological tissue with higher contrast and throughput.

- Spectroscopy, which measures the particles' interaction with electromagnetic radiation as a function of wavelength, is useful for some classes of nanoparticles to characterize concentration, size, and shape. Semiconductor quantum dots are fluorescent and metal nanoparticles exhibit surface plasmon absorbances, making both amenable to ultraviolet–visible spectroscopy.

- Infrared, nuclear magnetic resonance, and X-ray spectroscopy are also used with nanoparticles. Light scattering methods using laser light, X-rays, or neutron scattering are used to determine particle size, with each method suitable for different size ranges and particle compositions.

- Some miscellaneous methods are electrophoresis for surface charge, the Brunauer–Emmett–Teller method for surface area, and X-ray diffraction for crystal structure, as well as mass spectrometry for particle mass, and particle counters for particle number.[6] Chromatography, centrifugation, and filtration techniques can be used to separate nanoparticles by size or other physical properties before or during characterization.

## 1.6.2  Applications of Nano Characterization

### Product Verification

- Scanning electron microscope images of four samples of zinc oxide nanoparticles from different vendors, showing differences in size and shape. Characterization of nanomaterials by manufacturers and users is important in assessing the uniformity and reproducibility of their properties.

- Manufacturers and users of nanoparticles may perform characterization of their products for process control or verification and validation purposes. The properties of nanoparticles are sensitive to small variations in the processes used to synthesize and process them.

- Thus, nanoparticles prepared by seemingly identical processes must be characterized to determine if they are actually equivalent. Any material or dimensional property of a nanomaterial can be heterogeneous, and these can lead to heterogeneity in their functional properties.

- Generally, uniform collections are desired. It is advantageous to minimize heterogeneity during the initial synthesis, stabilization, and functionalization processes, rather than through downstream purification steps that decrease yield. Batch-to-batch reproducibility is also desirable. Unlike research-oriented nanometrology, industrial measurements emphasize reducing time, cost, and number of measured metrics, and must be performed under ambient conditions during a production process.

- Different applications have different tolerances for uniformity and reproducibility, and require different approaches to characterization. For example, nanocomposite materials may be tolerant of a broad distribution of nanoparticle properties.

- By contrast, characterization is especially important for nanomedicines, as their efficacy and safety depends strongly on critical properties such as particle size distribution, chemical composition, and the kinetics of drug loading and release. The development of standardized analytical methods for nanomedicines is in its early stages. However, standardized "assay cascades" have been developed to assist with this.

### Toxicology

- Nanotoxicology is the study of the toxic effects of nanoparticles on living organisms. Characterization of a nanoparticle's physical and chemical properties is important for ensuring the reproducibility of toxicology studies, and is also vital for studying how the physical and chemical properties of nanoparticles determine their biological effects.

- The properties of a nanoparticle, such as size distribution and agglomeration state, can change as a material is prepared and used in toxicology studies. This makes it important to measure them at different points in the experiment. The "as-received" or "as-generated" properties refer to the material's state when received from the manufacturer or synthesized in the laboratory.

- The "as-dosed" or "as-exposed" properties refer to its state when administered to the biological system. These may differ from the "as-received" state due to formation of aggregates and agglomerates if the material has been in powder form, the settling out of larger aggregates and agglomerates, or loss by adhesion to surfaces.

- The properties may again be different at the point of interaction with the organism's tissues due to biodistribution and physiological clearance mechanisms. At this stage, it is difficult to measure nanoparticle properties in situ without perturbing the system. Post mortem or histological examination provides a way to measure these changes in the material, although the tissue itself can interfere with the measurements.

**Exposure Assessment**

- Exposure assessment is a set of methods used to monitor contaminant release and exposures to workers. It is a vital step in mitigating the health and safety hazards of nanomaterials in workplaces where they are handled.
- For engineered nanoparticles, the assessment often involves use of both real-time instruments such as particle counters, which monitor the total number of particles in air (including both the nanoparticle of interest and other background particles), and filter-based occupational hygiene sampling methods that use electron microscopy and elemental analysis to identify the nanoparticle of interest.
- Personal sampling locates the samplers in the personal breathing zone of the worker, as close to the nose and mouth as possible and usually attached to a shirt collar. Area sampling is where samplers are placed at static locations.
- The NIOSH Nanomaterial Exposure Assessment Technique (NEAT 2.0) is a sampling strategy that can be used to determine exposure potential for engineered nanoparticles. The NEAT 2.0 approach uses two filter samples in the worker's personal breathing zone and as area samples. The first is used for elemental analysis.
- The second is used to gather morphologic data from electron microscopy, which can provide an order of magnitude evaluation of the contribution of the nanoparticle of interest to the elemental mass load, as well as a qualitative assessment of the particle size, degree of agglomeration, and whether the nanoparticle is free or contained within a matrix. Hazard identification and characterization can then be performed based on a holistic assessment of the integrated filter samples.
- In addition, field-portable direct reading instruments can be used for continuous recording of normal fluctuations in particle count, size distribution, and mass. By documenting the workers' activities, data-logged results can then be used to identify workplace tasks or practices that contribute to any increase or spikes in the counts.
- The data need to be carefully interpreted, as direct reading instruments will identify the real-time quantity of all nanoparticles including any incidental background particles such as may occur from motor exhaust, pump exhaust, heating vessels, and other sources. Evaluation of worker practices, ventilation efficacy, and other engineering exposure control systems and risk management strategies serve to allow for a comprehensive exposure assessment.
- To be effective, real-time particle counters should be able to detect a wide range of particle sizes, as nanoparticles may aggregate in the air. Adjacent work areas can be simultaneously tested to establish a background concentration. Not all instruments used to detect aerosols are suitable for monitoring occupational nanoparticle emissions because they may not be able to detect smaller particles, or may be too large or difficult to ship to a workplace.
- Some NIOSH methods developed for other chemicals can be used for off-line analysis of nanoparticles, including their morphology and geometry, elemental carbon content (relevant for carbon-based nanoparticles), and elemental analysis for several metals.
- Occupational exposure limits have not yet been developed for many of the large and growing number of engineered nanoparticles now being produced and used, as their hazards are not fully known. While mass-based metrics are traditionally used to characterize toxicological effects of exposure to air contaminants, it remains unclear which metrics are most important with regard to engineered nanoparticles.
- Animal and cell-culture studies have shown that size and shape may be two major factors in their toxicological effects. Surface area and surface chemistry also appear to be more important than mass concentration.
- The U.S. National Institute for Occupational Safety and Health has determined non-regulatory recommended exposure limits (RELs) of 1.0 $\mu g/m^3$ for carbon nanotubes and carbon nanofibers as background-corrected elemental carbon as an 8-hour time-weighted average (TWA) respirable mass concentration, and 300 $\mu g/m^3$ for ultrafine titanium dioxide as TWA concentrations for up to 10 hr/day during a 40-hour work week.

## 1.7 MESO STRUCTURES

- A meso compound or meso isomer is a non-optically active member of a set of stereoisomers, at least two of which are optically active. This means that despite containing two or more stereogenic centers, the molecule is not chiral. A meso compound is "superposable" on its mirror image (not to be confused with superimposable, as any two objects can be superimposed over one another regardless of whether they are the same). Two objects can be superposed if all aspects of the objects coincide and it does not produce a "(+)" or "(-)" reading when analyzed with a polarimeter.

- For example, tartaric acid can exist as any of three stereoisomers depicted below in a Fischer projection. Of the four colored pictures at the top of the diagram, the first two represent the meso compound (the 2R, 3S and 2S, 3R isomers are equivalent), followed by the optically active pair of levotartaric acid (L-(R,R)-(+)-tartaric acid) and dextrotartaric acid (D-(S,S)-(-)-tartaric acid).

- The meso compound is bisected by an internal plane of symmetry that is not present for the non-meso isomers (indicated by an X). That is, on reflecting the meso compound through a mirror plane perpendicular to the screen, the same stereochemistry is obtained; this is not the case for the non-meso tartaric acid, which generates the other enantiomer. The meso compound must not be confused with a 50:50 racemic mixture of the two optically-active compounds, although neither will rotate light in a polarimeter.

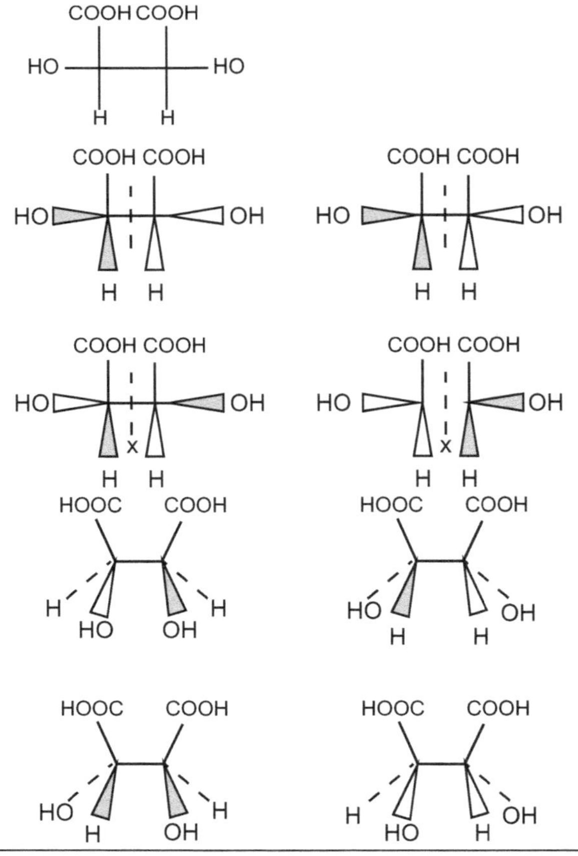

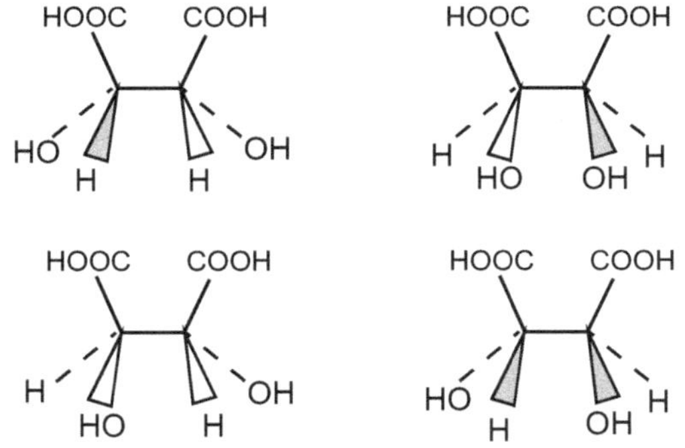

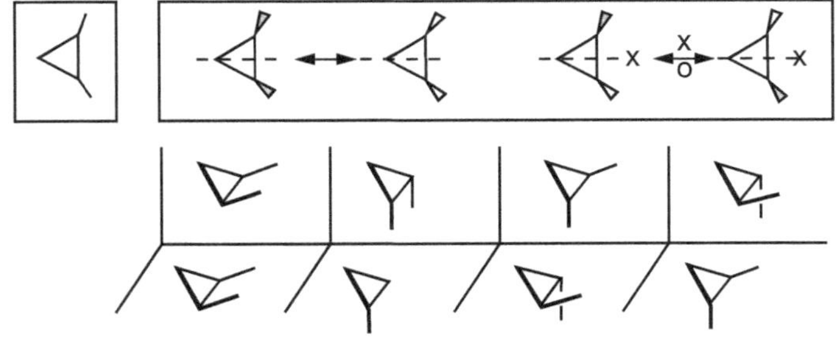

**Fig. 1.56**

It is a requirement for two of the stereocenters in a meso compound to have at least two substituents in common (although having this characteristic does not necessarily mean that the compound is meso). For example, in 2, 4-pentanediol, both the second and fourth carbon atoms, which are stereocenters, have all four substituents in common.

Since a meso isomer has a superposable mirror image, a compound with a total of $n$ chiral centers cannot attain the theoretical maximum of $2^n$ stereoisomers if one of the stereoisomers is meso.

**Cyclic Meso Compounds**

1,2-substituted cyclopropane has a meso cis-isomer (molecule has a mirror plane) and two trans-enantiomers:

**Fig. 1.57**

The two cis stereoisomers of 1, 2-substituted cyclohexanes behave like meso compounds at room temperature in most cases. At room temperature, most 1, 2-disubstituted cyclohexanes undergo rapid ring flipping (exceptions being rings with bulky substituents), and as a result, the two cis stereoisomers behave chemically identically with chiral reagents. At low temperatures, however, this is not the case, as the activation energy for the ring-flip cannot be overcome, and they therefore behave like enantiomers. Also noteworthy is the fact that when a cyclohexane undergoes a ring flip, the absolute configurations of the sterocenters do not change.

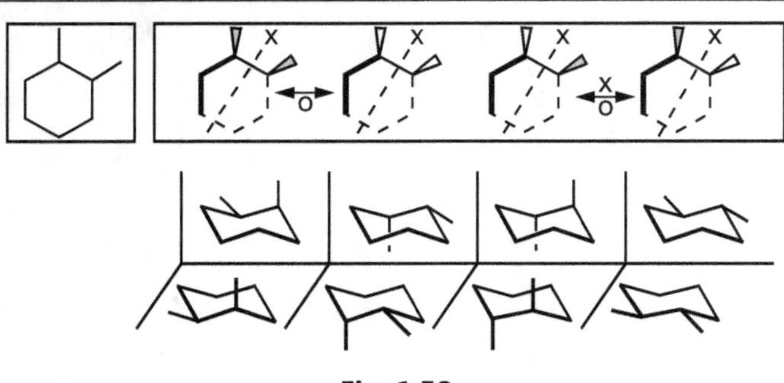

**Fig. 1.58**

## SUMMARY

- Basically nanoelectronics materials fabricated by the silicon, gallium and arsenic etc.

- Silicon nanoelectronics materials fabrication plays the important role in two important things. i.e quantum effect and single electron effect.

- Semiconductor band structures in general and especially for silicon are hard to describe with an analytical formula.

- The crystalline lattice is a periodic array of the atoms. When the solid is not crystalline, it is called amorphous.

- A crystal structure is composed of a unit cell, a set of atoms arranged in a particular way; which is periodically repeated in three dimensions on a lattice.

- Crystal defect, imperfection in the regular geometrical arrangement of the atoms in a crystalline solid.

- Point defects include self interstitial atoms, interstitial impurity atoms, substitution atoms and vacancies.

## EXERCISE

1. What do you understand by the crystal defects? Explain.

2. Explain the crystal growth and wafer fabrication.

3. What are the crystal planes and its orientation?

4. Recognize the basic steps of CMOS.

5. Discuss I–V/C-V characterization.

6. Explain the temperature dependent characterization.

7. Write a short note on the CMOS technology.

8. State and explain the construction of mass field effect transistor.

9. Explain the water fabrication techniques.

10. State the working principal of N and P-MOSFET transistor.

◈ ◈ ◈

# BASICS OF QUANTUM MECHANICS

## 2.1 INTRODUCTION

- Quantum mechanics (QM; also known as quantum physics, quantum theory, the wave mechanical model, or matrix mechanics), including quantum field theory, is a fundamental theory in physics which describes nature at the smallest scales of atoms and subatomic particles.

- Classical physics, the description of physics existing before the formulation of the theory of relativity and of quantum mechanics, describes nature at ordinary (macroscopic) scale. Most theories in classical physics can be derived from quantum mechanics as an approximation valid at large (macroscopic) scale.

- Quantum mechanics differs from classical physics in that energy, momentum, angular momentum, and other quantities of a bound system are restricted to discrete values (quantization), objects have characteristics of both particles and waves (wave-particle duality), and there are limits to the precision with which quantities can be measured (the uncertainty principle).

- Quantum mechanics gradually arose from theories to explain observations which could not be reconciled with classical physics, such as Max Planck's solution in 1900 to the black-body radiation problem, and from the correspondence between energy and frequency in Albert Einstein's 1905 paper which explained the photoelectric effect. Early quantum theory was profoundly re-conceived in the mid-1920s by Erwin Schrödinger, Werner Heisenberg, Max Born and others.

- The modern theory is formulated in various specially developed mathematical formalisms. In one of them, a mathematical function, the wave function, provides information about the probability amplitude of position, momentum, and other physical properties of a particle.

## 2.2 SCHRODINGER EQUATION

- The Schrodinger equation plays the role of Newton's laws and conservation of energy in classical mechanics - i.e., it predicts the future behavior of a dynamic system. It is a wave equation in terms of the wavefunction which predicts analytically and precisely the probability of events or outcome. The detailed outcome is not strictly determined, but given a large number of events, the Schrodinger equation will predict the distribution of results.

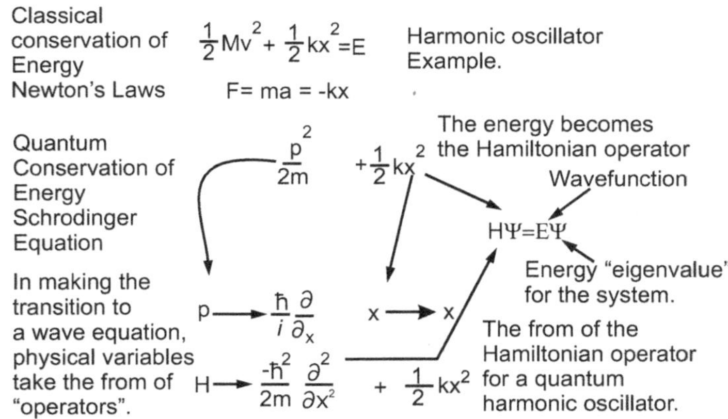



- The kinetic and potential energies are transformed into the Hamiltonian which acts upon the wave function to generate the evolution of the wave function in time and space. The Schrodinger equation gives the quantized energies of the system and gives the form of the wave function so that other properties may be calculated.

- At the beginning of the twentieth century, experimental evidence suggested that atomic particles were also wave-like in nature. For example, electrons were found to give diffraction patterns when passed through a double slit in a similar way to light waves. Therefore, it was reasonable to assume that a wave equation could explain the behaviour of atomic particles.

- Schrodinger was the first person to write down such a wave equation. Much discussion then centred on what the equation meant. The eigenvalues of the wave equation were shown to be equal to the energy levels of the quantum mechanical system, and the best test of the equation was when it was used to solve for the energy levels of the Hydrogen atom, and the energy levels were found to be in accord with Rydberg's Law.

- It was initially much less obvious what the wavefunction of the equation was. After much debate, the wavefunction is now accepted to be a probability distribution. The Schrodinger equation is used to find the allowed energy levels of quantum mechanical systems (such as atoms, or transistors). The associated wavefunction gives the probability of finding the particle at a certain position.

Answered by: Ian Taylor, Ph.D., Theoretical Physics (Cambridge), PhD (Durham), UK

- The Shrodinger equation is:

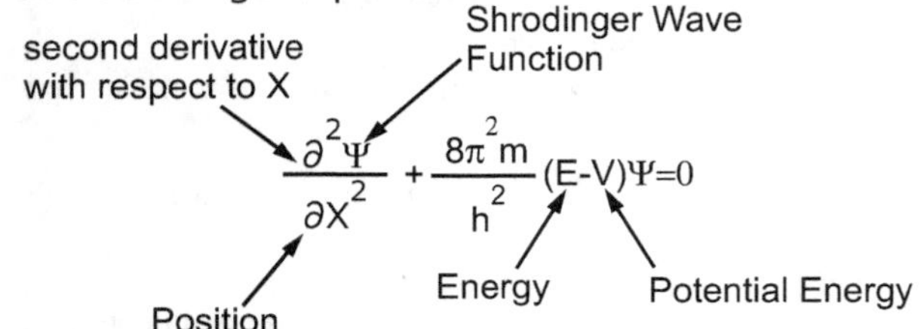


$$\frac{\partial^2 \Psi}{\partial x^2} + \frac{8\pi^2 m}{h^2}(E-V)\Psi = 0$$

- The solution to this equation is a wave that describes the quantum aspects of a system. However, physically interpreting the wave is one of the main philosophical problems of quantum mechanics.

- The solution to the equation is based on the method of Eigen Values devised by Fourier. This is where any mathematical function is expressed as the sum of an infinite series of other periodic functions. The trick is to find the correct functions that have the right amplitudes so that when added together by superposition they give the desired solution.

- So, the solution to Schrondinger's equation, the wave function for the system, was replaced by the wave functions of the individual series, natural harmonics of each other, an infinite series. Shrodinger has discovered that the replacement waves described the individual states of the quantum system and their amplitudes gave the relative importance of that state to the whole system. Schrodinger's equation shows all of the wave like properties of matter and was one of greatest achievements of 20th century science. It is used in physics and most of chemistry to deal with problems about the atomic structure of matter.

- It is an extremely powerful mathematical tool and the whole basis of wave mechanics.

- Answered by: Simon Hooks, Physics A-Level Student, Gosport, UK The Schrodinger equation is the name of the basic non-relativistic wave equation used in one version of quantum mechanics to describe the behaviour of a particle in a field of force. There is the time dependant equation used for describing progressive waves, applicable to the motion of free particles. And the time independent form of this equation used for describing standing waves.

- Schrodinger's time-independent equation can be solved analytically for a number of simple systems. The

time-dependant equation is of the first order in time but of the second order with respect to the co-ordinates, hence it is not consistent with relativity. The solutions for bound systems give three quantum numbers, corresponding to three co-ordinates, and an approximate relativistic correction is possible by including fourth spin quantum number.

## 2.3 DENSITY OF STATES

- The quantity which represents the number of allowed electron or hole states per volume at a given energy is called density of states. Many of the bulk material properties like paramagnetic susceptibility, specific heat depend on density of states.

- Density of state calculation is handy in determining the energy band spacing in semiconductors. It also helps in determining general distribution of states as a function of energy.

- The density of states for a wave in one dimension is represented by the equation

$$D(\omega) = \frac{L}{\pi}\frac{1}{v_s}$$

Here, $L$ is the system length, $v_s$ represent speed of sound, and $w$ represents frequency of wave.

- The density of states for a wave in two dimensions is,

$$D(\omega) = \frac{L^2}{\pi}\frac{\omega}{v_s^2}$$

- As we consider electrons present in metal the density of states arises from wave nature of electron in the particle in a box like setting. Here the density of state is given by the equation

$$D(E) = \frac{4\pi (2m)^{3/2}}{h^3}\sqrt{E}$$

- For a photon we have the relationship between wavelength and its energy

$$E = \frac{hc}{\lambda}$$

Here, $h$ represents the Planck's constant, $c$ represents speed of light, and $l$ represents wavelength of photon. Thus, the density of its state for photon is given by the relation,

$$D(E) = \frac{8\pi}{(hc)^3}E^2$$

- Value of density of states conveys lots of meaning. A very high value of density of sates signifies that there is large number of states available for occupation. Similarly, a zero value for density of state represent that no state can be occupied at that particular energy level. The calculation of density of state is found to be easier when symmetry of the system is higher.

## 2.4  DEGENERACY

- In quantum mechanics, an energy level is degenerate if it corresponds to two or more different measurable states of a quantum system. Conversely, two or more different states of a quantum mechanical system are said to be degenerate if they give the same value of energy upon measurement.

- The number of different states corresponding to a particular energy level is known as the degree of degeneracy of the level. It is represented mathematically by the Hamiltonian for the system having more than one linearly independent eigenstate with the same energy eigenvalue. When this is the case, energy alone is not enough to characterize what state the system is in, and other quantum numbers are needed to characterize the exact state when distinction is desired. In classical mechanics, this can be understood in terms of different possible trajectories corresponding to the same energy.

- Degeneracy plays a fundamental role in quantum statistical mechanics. For an N-particle system in three dimensions, a single energy level may correspond to several different wave functions or energy states. These degenerate states at the same level are all equally probable of being filled. The number of such states gives the degeneracy of a particular energy level.

- Degenerate is used in quantum mechanics to mean 'of equal energy.' It usually refers to electron energy levels or sublevels.

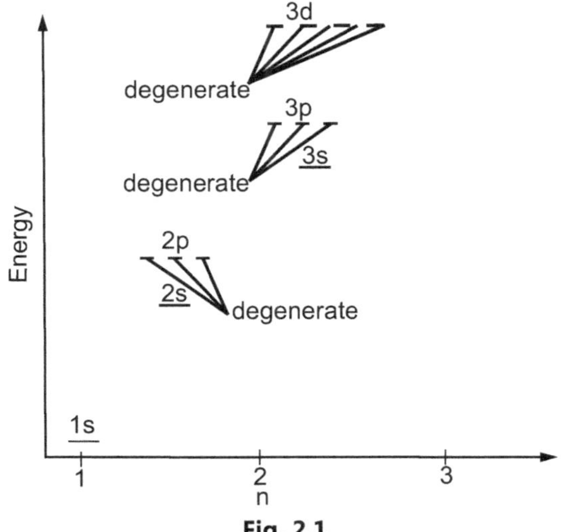


**Fig. 2.1**

- For example, orbitals in the 2p sublevel are degenerate - in other words the $2p_x$, $2p_y$, and $2p_z$ orbitals are equal in energy, as shown in the Fig. 2.1.

- Likewise, at a higher energy than 2p, the $3p_x$, $3p_y$, and $3p_z$ orbitals are degenerate. And, at a still higher energy, the $3d_{xy}$, $3d_{xz}$, $3d_{yz}$, $3d_x{}^2 - y^2$, and $3d_z{}^2$ are degenerate.

- The number of different states of equal energy is called the degree of degeneracy or just degeneracy.

- The degeneracy of p orbitals is 3; the degeneracy of d orbitals is 5; the degeneracy of f orbitals is 7.

- We can also compare electron energies. In the following Fig. 2.2 of hydrogen atom energy levels, the electrons are degenerate.

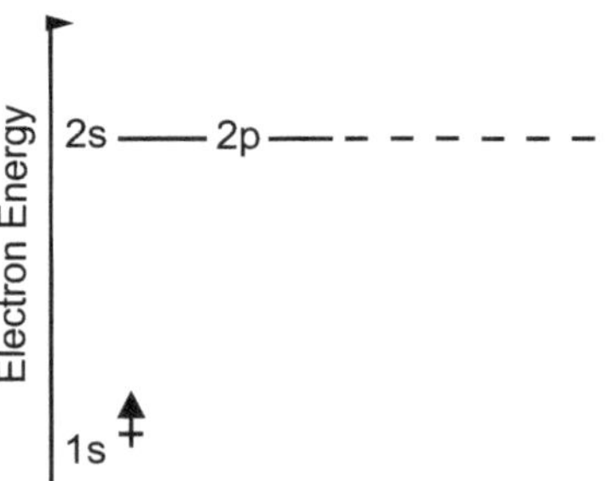


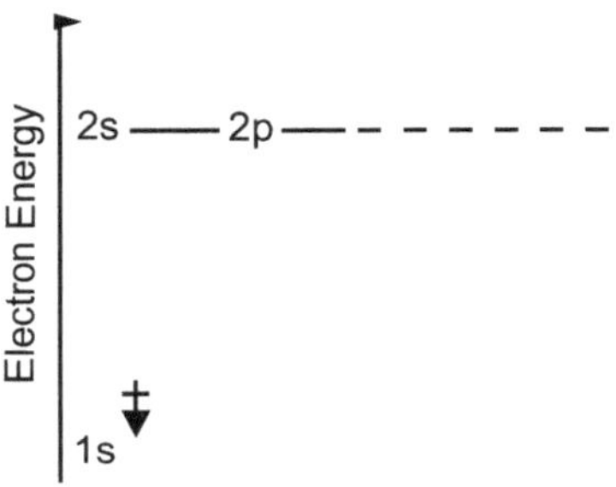


**Fig. 2.2**

- One of the electrons is spin-up and the other is spin-down. In a non-uniform magnetic field, different spins respond differently; electrons with different spin orientations would no longer be degenerate - they would have slightly different amounts of energy. This is how electron spin was first detected in 1922 by the Stern-Gerlach experiment.

- Notice also in these two diagrams that the 2s and 2p sublevels are degenerate. This is the case for hydrogenic atoms and ions - i.e. those with only one electron.

## 2.5  BAND THEORY OF SOLIDS

- A useful way to visualize the difference between conductors, insulators and semiconductors is to plot the available energies for electrons in the materials. Instead of having discrete energies as in the case of free atoms, the available energy states form bands. Crucial to the conduction process is whether or not there are electrons in the conduction band.

- In insulators the electrons in the valence band are separated by a large gap from the conduction band, in conductors like metals the valence band overlaps the conduction band, and in semiconductors there is a small enough gap between the valence and conduction bands that thermal or other excitations can bridge the gap. With such a small gap, the presence of a small percentage of a doping material can increase conductivity dramatically.

- An important parameter in the band theory is the Fermi level, the top of the available electron energy levels at low temperatures. The position of the Fermi level with the relation to the conduction band is a crucial factor in determining electrical properties.

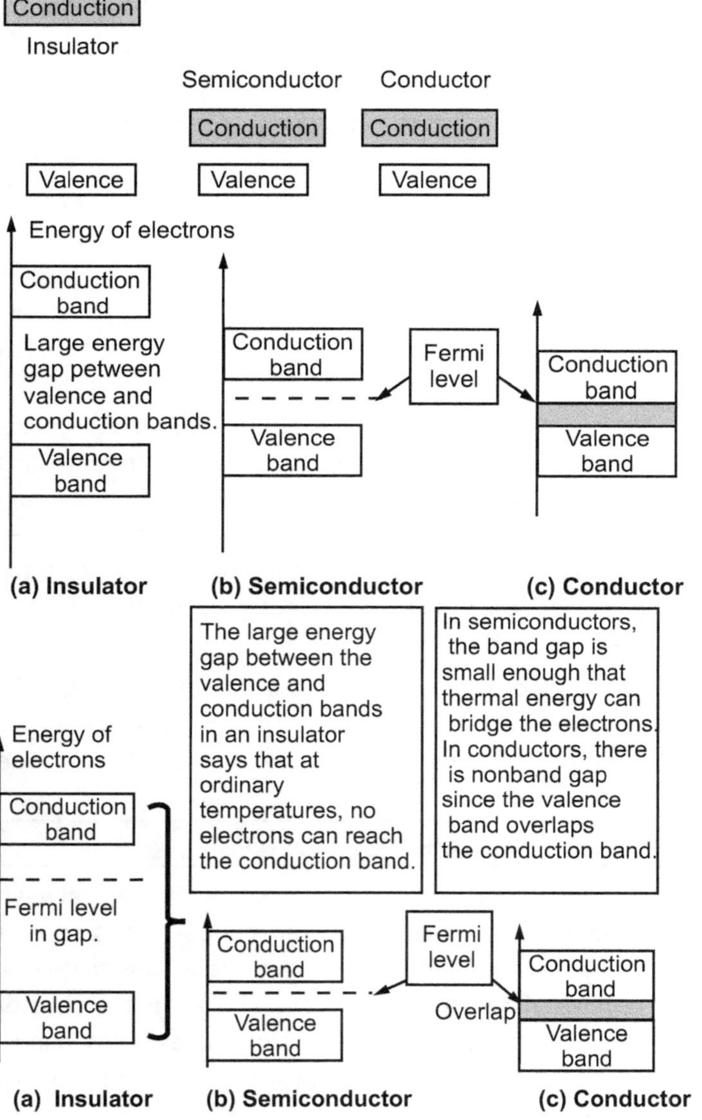


**Fig. 2.3**

### 2.5.1 Energy Bands in Conductor

- In terms of the band theory of solids, metals are unique as good conductors of electricity. This can be seen to be a result of their valence electrons being essentially free. In the band theory, this is depicted as an overlap of the valence band and the conduction band so that at least a fraction of the valence electrons can move through the material.

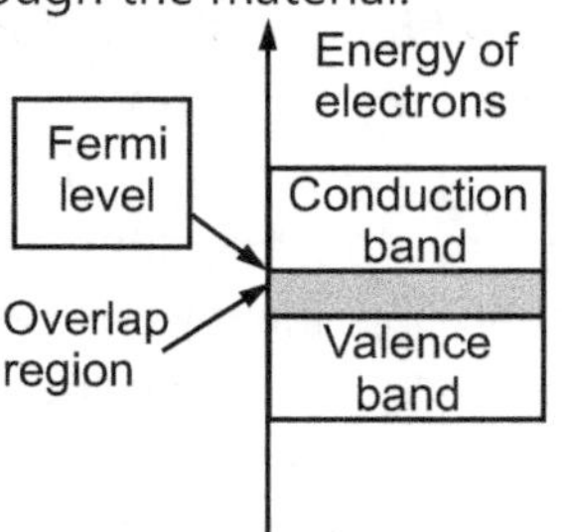


**Fig. 2.4**

### 2.5.2 Energy Bands in Insulator

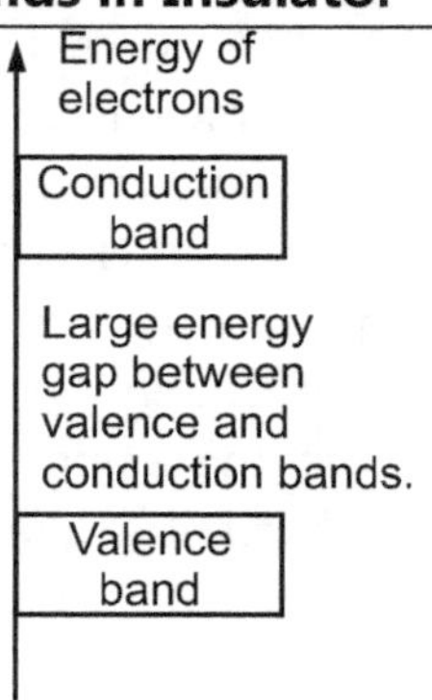


**Fig. 2.5**

- Most solid substances are insulators, and in terms of the band theory of solids this implies that there is a large forbidden gap between the energies of the valence electrons and the energy at which the electrons can move freely through the material (the conduction band).

- Glass is an insulating material which may be transparent to visible light for reasons closely correlated with its nature as an electrical insulator. The visible light photons do not have enough quantum energy to bridge the band gap and get the electrons up to an available energy level in the conduction band. The visible properties of glass can also give some insight into the effects of "doping" on the properties of solids.

- A very small percentage of impurity atoms in the glass can give it color by providing specific available energy levels which absorb certain colors of visible light. The ruby mineral (corundum) is aluminum oxide with a small amount (about 0.05%) of chromium which gives it its characteristic pink or red color by absorbing green and blue light.

- While the doping of insulators can dramatically change their optical properties, it is not enough to overcome the large band gap to make them good conductors of electricity. However, the doping of semiconductors has a much more dramatic effect on their electrical conductivity and is the basis for solid state electronics.

### 2.5.3 Energy Bands in Semiconductor

- For intrinsic semiconductors like silicon and germanium, the Fermi level is essentially halfway between the valence and conduction bands. Although no conduction occurs at 0 K, at higher temperatures a finite number of electrons can reach the conduction band and provide some current. In doped semiconductors, extra energy levels are added.

- The increase in conductivity with temperature can be modeled in terms of the Fermi function, which allows one to calculate the population of the conduction band.

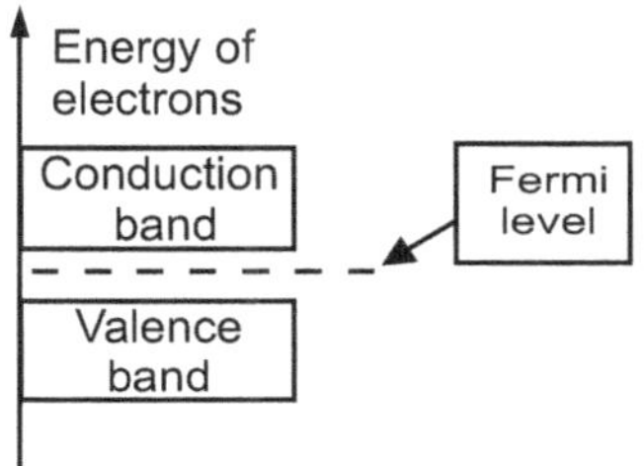


**Fig. 2.6**

### 2.5.4 Energy Bands in Silicon

- At finite temperatures, the number of electrons which reach the conduction band and contribute to current can be modeled by the Fermi function. That current is small compared to that in doped semiconductors under the same conditions.

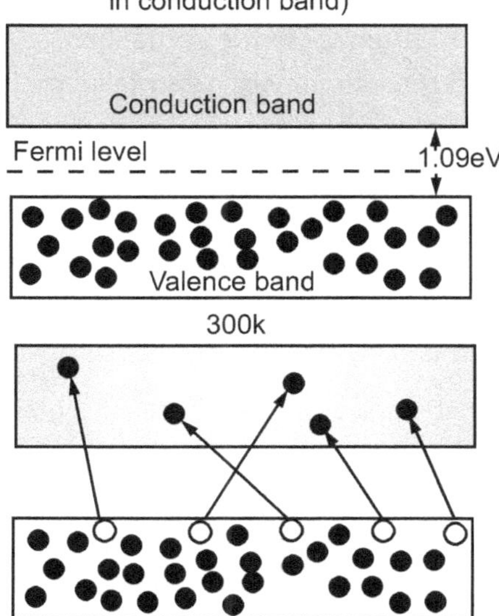


**Fig. 2.7**

### 2.5.5 Energy Bands in Germanium

- At finite temperatures, the number of electrons which reach the conduction band and contribute to current can be modeled by the Fermi function. That current is small compared to that in doped semiconductors under the same conditions.

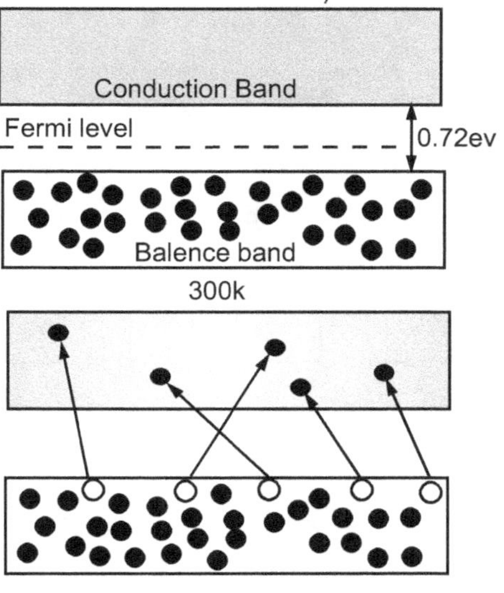


**Fig. 2.8**

## 2.6 KRONIG-PENNY MODEL

- According to quantum free electron theory of metals, a conduction electron in a metal experiences constant (or zero) potential and free to move inside the crystal but will not come out of the metal because an infinite potential exists at the surface. This theory successfully explains electrical conductivity, specific heat, thermionic emission and paramagnetism.

- This theory is fails to explain many other physical properties, for example:

  1. It fails to explain the difference between conductors, insulators and semiconductors,

  2. Positive Hall coefficient of metals and

  3. lower conductivity of divalent metals than monovalent metals.

- To overcome the above problems, the periodic potentials due to the positive ions in a metal have been considered. shown in Fig. 2.9 (a), if an electron moves through these ions, it experiences varying potentials. The potential of an electron at the positive ion site is zero and is maximum in between two ions. The potential experienced by an electron, when it passes along a line through the positive ions is as shown in Fig. 2.9 (b).

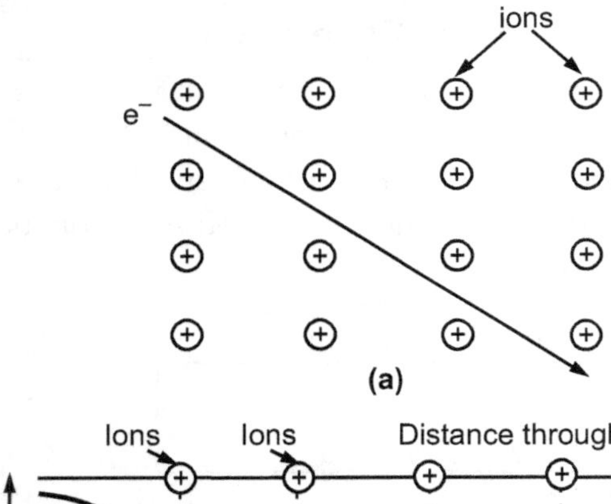



(a)

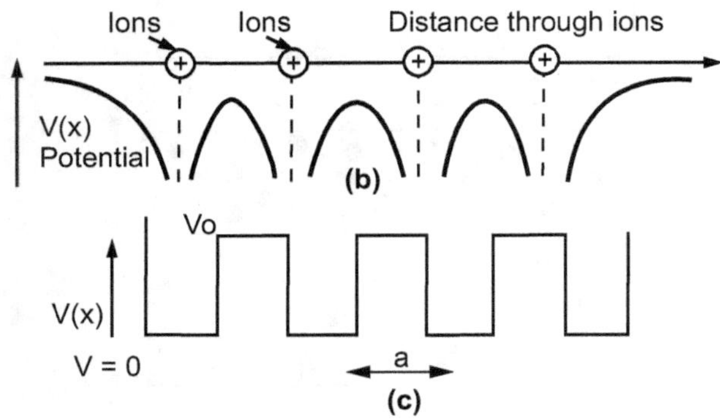





**Fig. 2.9**

- It is not easy to solve Schrödinger's equation with these potentials. So, Kronig and Penney approximated these potentials inside the crystal to the shape of rectangular steps as shown in Fig. 2.9 (c). This model is called Kronig-Penney model of potentials.

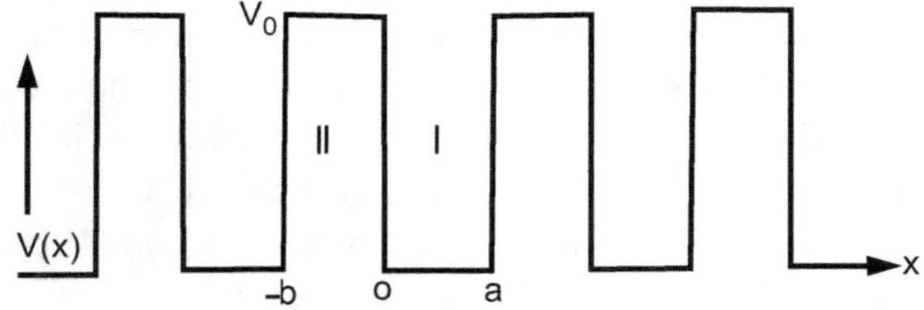



**Fig. 2.10**

- The energies of electrons can be known by solving Schrödinger's wave equation in such a lattice. The Schrodinger time-independent wave equation for the motion of an electron along X-direction is given by:

$$\frac{d^2\psi}{dx^2} + \frac{2m}{h^2}[E - V(x)]\,\psi = 0 \qquad \ldots(2.1)$$

- The energies and wave functions of electrons associated with this model can be calculated by solving time-independent one-dimensional Schrödinger's wave equations for the two regions I and II as shown in Fig. 2.10.

- The Schrodinger's equations are:

$$\frac{d^2\psi}{dx^2} + \frac{2m}{h^2}\,E\psi = 0 \text{ for } 0 < x < a \qquad \ldots(2.2)$$

$$\frac{d^2\psi}{dx^2} + \frac{2m}{h^2}[E - V_o]\,\psi = 0 \text{ for } -b < x < 0 \qquad \ldots(2.3)$$

We define two real quantities (say) $\alpha$ and $\beta$ such that:

$$\alpha^2 = \frac{2mE}{h^2}$$

and

$$\beta^2 = \frac{2m}{h^2}(V_0 - E) \qquad \ldots(2.4)$$

Hence, Equations (5.2) and (5.3) becomes:

$$\frac{d^2\psi}{dx^2} + \alpha^2\psi = 0 \text{ for } 0 < x < a$$

$$\frac{d^2\psi}{dx^2} - \beta^2\psi = 0 \text{ for } -b < x < 0$$

- According to Bloch's theorem, the wavefunction solution of the Schrödinger equation when the potential is periodic and to make sure the function $u(x)$ is also continuous and smooth, can be written as:

$$\psi(x) = e^{ikx}\,u(x)$$

Where $u(x)$ is a periodic function which satisfies $u(x + a) = u(x)$.

- Using Bloch theorem and all the boundary conditions for the continuity of the wave function the solution of Schrodinger wave equation obtained as

$$P\frac{\sin \alpha a}{\alpha a} + \cos\alpha a = \cos Ka$$

where

$$P = \frac{mV_o ab}{h^2}$$

- This equation shows the relation between the energy (through $\alpha$) and the wave-vector, $k$, and as you can see, since the left hand side of the equation can only range from −1 to 1 then there are some limits on the values that $\alpha$ (and thus, the energy) can take, that is, at some ranges of values of the energy, there is no solution according to these equation, and thus, the system will not have those energies: energy gaps. These are the so-called band-gaps, which can be shown to exist in *any* shape of periodic potential (not just delta or square barriers).

- Conversly if suppose the effect of potential barrier dominate i.e., if P is large, the resultant wave obtained in terms of shows a stepper variation in the region lies between +1 to −1. This results in the decrease of allowed energy and increase of forbidden energy gap. Thus at extremities,

**Case 1 :** When $P \to \infty$, the allowed energy states are compressed to a line spectrum.

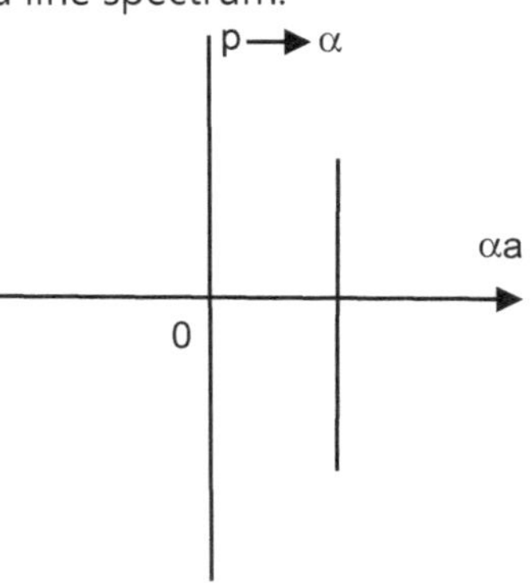

Fig. 2.11

**Case 2 :** When $P \to 0$ the energy band is broadened and it is quasi continuous.

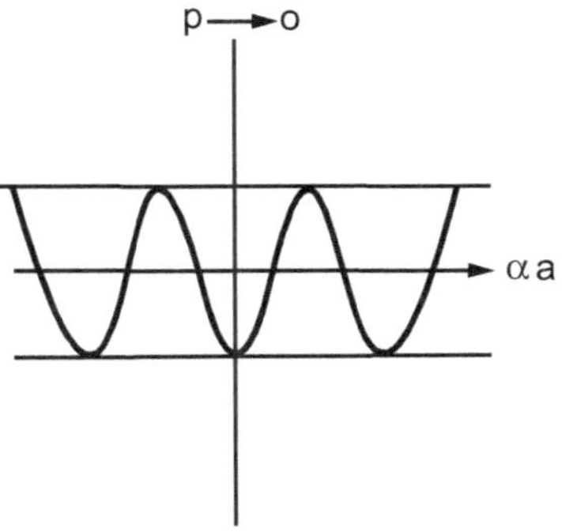

Fig. 2.12

## 2.7 BRILLOUIN ZONES

- In mathematics and solid state physics, the first Brillouin zone is a uniquely defined primitive cell in reciprocal space. In the same way the Bravais lattice is divided up into Wigner–Seitz cells in the real lattice, the reciprocal lattice is broken up into Brillouin zones.

- The boundaries of this cell are given by planes related to points on the reciprocal lattice. The importance of the Brillouin zone stems from the Bloch wave description of waves in a periodic medium, in which it is found that the solutions can be completely characterized by their behavior in a single Brillouin zone.

- The first Brillouin zone is the locus of points in reciprocal space that are closer to the origin of the reciprocal lattice than they are to any other reciprocal lattice points (see the derivation of the Wigner-Seitz cell). Another definition is as the set of points in $k$-space that can be reached from the origin without crossing any Bragg plane. Equivalently, this is the Voronoi cell around the origin of the reciprocal lattice.

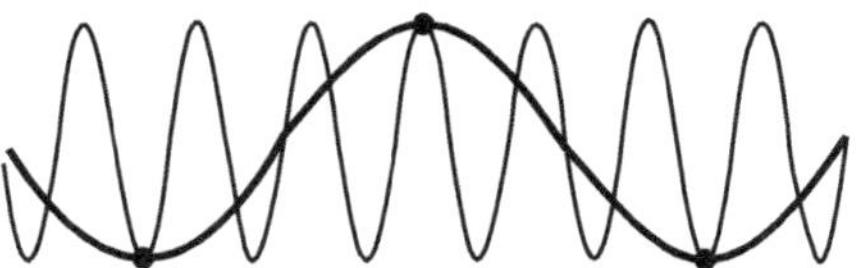

Fig. 2.13

- k-vectors exceeding the first Brillouin zone (red) do not carry any more information than their counterparts (black) in the first Brillouin zone. k at the Brilliouin zone edge is the spatial Nyquist frequency of waves in the lattice, because it corresponds to a half-wavelength equal to the inter-atomic lattice spacing $a$. See also Aliasing § Sampling sinusoidal functions for more on the equivalence of k-vectors.

- There are also second, third, *etc.*, Brillouin zones, corresponding to a sequence of disjoint regions (all with the same volume) at increasing distances from the origin, but these are used less frequently. As a result, the first Brillouin zone is often called simply the Brillouin zone. In general, the n-th Brillouin zone consists of the set of points that can be reached from the origin by crossing exactly n – 1 distinct Bragg planes.

- A related concept is that of the irreducible Brillouin zone, which is the first Brillouin zone reduced by all of the symmetries in the point group of the lattice (point group of the crystal).

- The concept of a Brillouin zone was developed by Léon Brillouin (1889–1969), a French physicist.

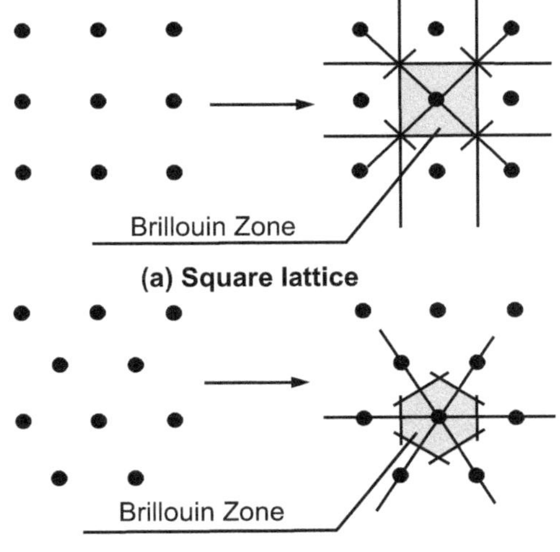


Fig. 2.14

- Fig. 2.14 The reciprocal lattices (dots) and corresponding first Brillouin zones of (a) square lattice and (b) hexagonal lattice.

- A Brillouin zone is a particular choice of the unit cell of the reciprocal lattice. It is defined as the Wigner-Seitz cell (also called Dirichlet or Voronoi domain of influence) of the reciprocal lattice. It is constructed as the set of points enclosed by the Bragg planes, the planes perpendicular to a connection line from the origin to each lattice point and passing through the midpoint shown in Fig. 2.14.

- Alternatively, it is defined as the set of points closer to the origin than to any other reciprocal lattice point. The whole reciprocal space may be covered without overlap with copies of such a Brillouin zone.

- For high-symmetry lattices one introduces sometimes the notion of *n-th Brillouin Zone*. This is the set of points one reaches with a straight line from the origin and passing through *n*-1 Bragg planes. In this terminology, the Brillouin zone defined above is the first Brillouin zone. The *n*-th Brillouin zone is a shell around lower Brillouin zones and its shape becomes for higher values of *n* rapidly rather complicated shown in Fig. 2.15.

- Vectors in the Brillouin zone or on its boundary characterize states in a system with lattice periodicity, *e.g.* phonon or electron states. States are non-equivalent if they belong to different vectors in a unit cell of the reciprocal lattice. This is not necessarily the Brillouin zone. Especially, for low-symmetry systems Brillouin zones are sometimes difficult to visualize, and another choice of unit cell may be useful, *e.g.* the parallelepiped spanned by the basis vectors.

- The Brillouin zone boundary consists of pieces of Bragg planes. Since two points on the boundary may differ by a reciprocal lattice vector, only a part of the boundary may be used for characterization of states. There is no simple rule for that choice. Dispersion curves at the zone boundary have either a maximum or minimum there or they merge with other curves giving rise to a change of degeneracy.

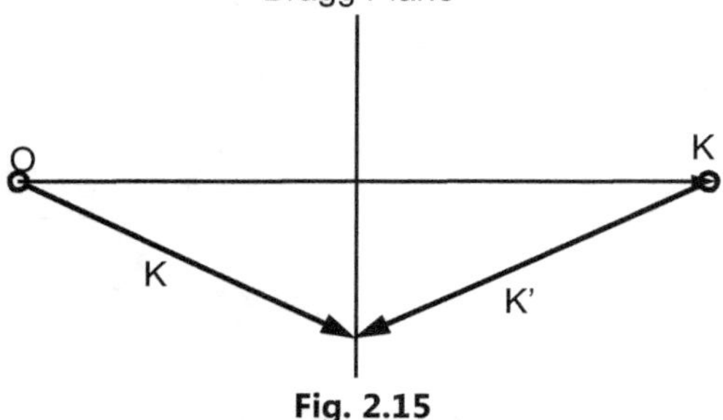


**Fig. 2.15**

- Bragg plane perpendicular to and going through the midpoint of the line connecting the origin and K. Particles in the crystal potential are elastically scattered from wave vector k to wave vector k'=k-K. The tips are in the Bragg plane.

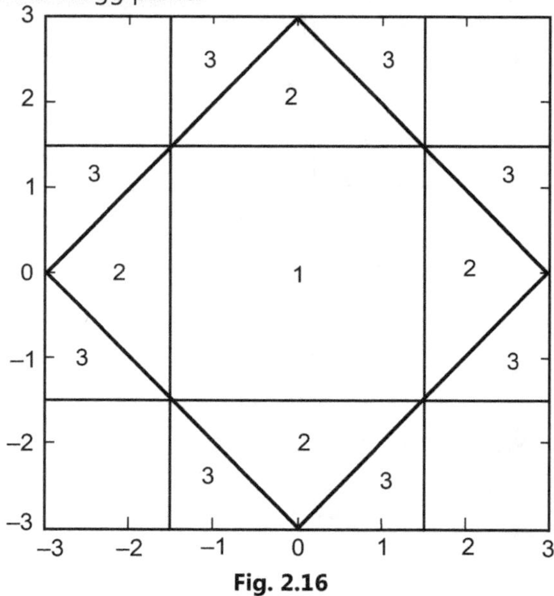

**Fig. 2.16**

- Up to the 3rd Brillouin zone for a two-dimensional square lattice; red lines are Bragg planes.

## SUMMARY

- The Schrodinger equation plays the role of Newton's laws and the conservation of energy in classical mechanics.

- The Schrodinger equation is

$$\frac{\partial^2 \psi}{\partial x^2} + \frac{8\pi^2 m}{h^2} = (E - V)\, \psi = 0$$

- An important parameter in the band theory is the, Fermi level the top of the available electron energy levels at low temperatures. The position of the Fermi level with the relation to the conduction band is a factor in determining electrical properties.

- The first Brillouin zone is a uniquely defined primitive cell in reciprocal space. Bravais lattice is divided up into Winger-Seitz cells in the real lattice, the reciprocal lattice is broken up into Brillouin Zones.

## EXERCISE

1. Explain Quantum Mechanics in details.
2. Describe Schrodinger equation.
3. Explain Density of States.
4. Describe Particle in a box Concepts.
5. What is Degeneracy.
6. Draw and explain Band Theory of Solids.
7. Explain Kronig-Penny Model.
8. Describe Brillouin Zones

# MOS SCALING THEORY

## 3.1 INTRODUCTION

- Over the past three decades, CMOS technology scaling has been a primary driver of the electronics industry and has provided a path toward both denser and faster integration. The transistors manufactured today are 20 times faster and occupy less than 1% of the area of those built 20 years ago.
- The number of devices per chip and the system performance has been improving exponentially over the last two decades. As the channel length is reduced, the performance improves, the power per switching event decreases, and the density improves. But the power density, total circuits per chip, and the total chip power consumption has been increasing.
- The need for more performance and integration has accelerated the scaling trends in almost every device parameter, such as lithography, effective channel length, gate dielectric thickness, supply voltage, device leakage, etc.
- Some of these parameters are approaching fundamental limits, and alternatives to the existing material and structures may need to be identified in order to continue scaling.

## 3.2 CMOS SCALING THEORY

- During the early 1970s, both Mead and Dennard noted that the basic MOS transistor structure could be scaled to smaller physical dimensions. The scaling theory developed by Mead and Dennard allows a "photocopy reduction" approach to feature size reduction in CMOS technology, and while the dimensions shrink, scaling theory causes the field strengths in the MOS transistor to remain the same across different process generations.
- Thus, the "original" form of scaling theory is constant field scaling. Constant field scaling requires a reduction of the power supply voltage with each technology generation. In the 1980s, CMOS adopted the 5V power supply, which was compatible with the power supply of bipolar TTL logic.
- Constant field scaling was replaced with constant voltage scaling, and instead of remaining constant, the fields inside the device increased from generation to generation until the early 1990s, when excessive power dissipation and heating, gate dielectrics TDDB and channel hot carrier aging caused serious problems with the increasing electric field. As a result, constant field scaling was applied to technology scaling in the 1990s.

**Moore's Law**

- It was the realization of scaling theory and its usage in practice which has made possible the better-known "Moore's Law." Moore's Law is a phenomenological observation that the number of transistors on integrated circuits doubles every two years, as shown in Fig. 3.1.
- It is intuitive that Moore's Law cannot be sustained forever. However, predictions of size reduction limits due to material or design constraints, or even the pace of size reduction, have proven to elude the most insightful scientists.

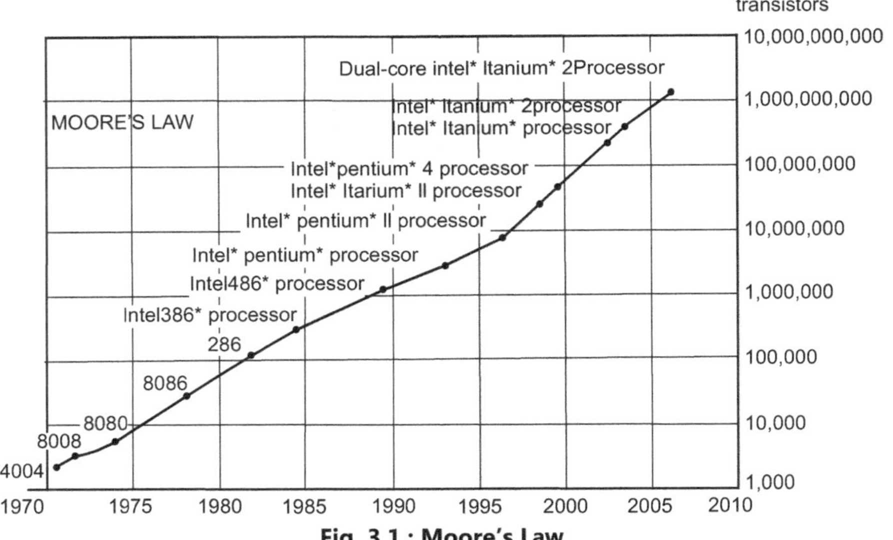


**Fig. 3.1 : Moore's Law**

**Scaling to its Limits**

- There does not seem to be any fundamental physical limitation that would prevent Moore's Law from characterizing the trends of integrated circuits.

- However, sustaining this rate of progress is not a straightforward achievement. Figure 2 shows the trends of power supply voltage, threshold voltage, and gate oxide thickness versus channel length for high performance CMOS logic technologies.

- Sub-threshold non-scaling and standby power limitations bound the threshold voltage to a minimum of 0.2 V at the operating temperature. Thus, a significant reduction in performance gains is predicted below 1.5 V due to the fact that the threshold voltage decreases more slowly than the historical trend, leading to more aggressive device designs at higher electric fields.

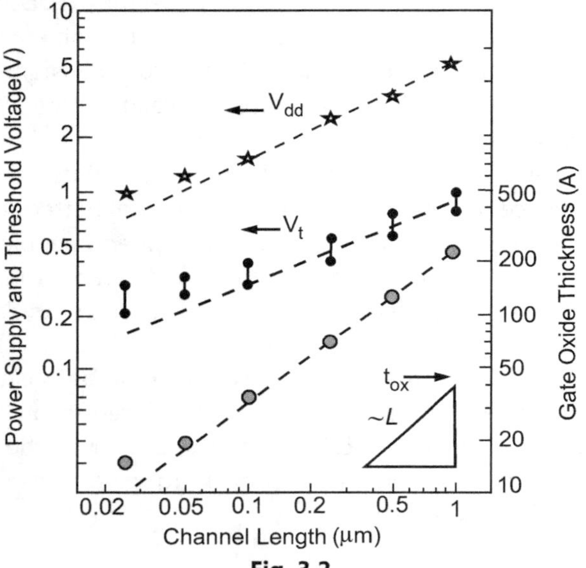


**Fig. 3.2**

- Fig. 3.2 Trends of power supply voltage Vdd, threshold voltage Vth, and gate oxide thickness tox, versus channel length for CMOS logic technologies.

- Further technology scaling requires major changes in many areas, including:
  1. Improved lithography techniques and non-optical exposure technologies;
  2. Improved transistor design to achieve higher performance with smaller dimensions;
  3. Migration from current bulk CMOS devices to novel materials and structures, including silicon-on- insulator, strained Si and novel dielectric materials;
  4. Circuit sensitivity to soft errors from radiation;
  5. Smaller wiring for on-chip interconnection of the circuits;
  6. Stable circuits;

  7. More productive design automation tools.

In addition, packaging technology needs to progress at a rate consistent with on- going CMOS technology scaling at sustainable cost/performance levels. This requires advances in I/O density, bandwidth, power distribution, and heat extraction. System architecture will also be required to maximize the performance gains achieved in advanced CMOS and packaging technologies.

**Scaling Impact on Circuit Performance**

- Transistor scaling is the primary factor in achieving high-performance microprocessors and memories.

- Each 30% reduction in CMOS IC technology node scaling has
  1. Reduced the gate delay by 30% allowing an increase in maximum clock frequency of 43%;
  2. Doubled the device density;
  3. Reduced the parasitic capacitance by 30%; and
  4. Reduced energy and active power per transition by 65% and 50%, respectively.

Fig. 3.3 shows CMOS performance, power density and circuit density trends, indicating a linear circuit performance as a result of technology scaling.

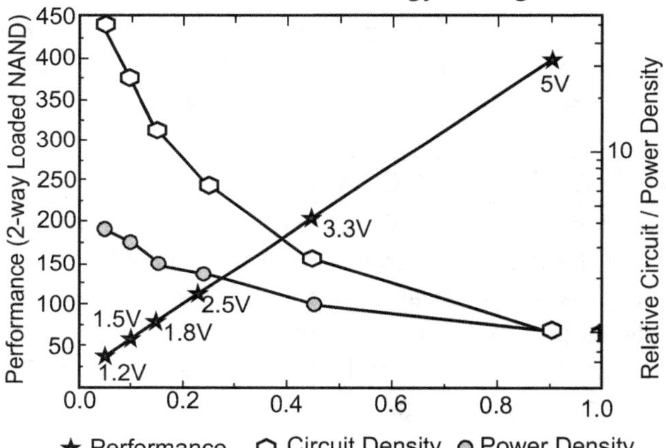


**Fig. 3.3 : MOS performance, power density and circuit density trends**

## 3.3 SHRINK-DOWN APPROACHES

- The term shrink (sometimes optical shrink or process shrink) refers to the scaling of Metal-Oxide-Semiconductor (MOS) devices. The act of shrinking a die is to create a somewhat identical circuit using a more advanced fabrication process, usually involving an advance of lithographic nodes. This reduces overall costs for a chip company, as the absence of major architectural changes to the processor lowers research and development costs, while at the same time allowing more processor dies to be manufactured on the same piece of silicon wafer, resulting in less cost per product sold.

- Die shrinks are the key to improving price/performance at semiconductor companies such as Samsung, Intel, TSMC, and SK Hynix, and fabless manufacturers such as AMD (including the former ATI), NVIDIA and MediaTek.

- Examples in the 2000s include the downscaling of the PlayStation 2's Emotion Engine processor from Sony and Toshiba (from 180 nm CMOS in 2000 to 90 nm CMOS in 2003), the codenamed Cedar Mill Pentium 4 processors (from 90 nm CMOS to 65 nm CMOS) and Penryn Core 2 processors (from 65 nm CMOS to 45 nm CMOS), the codenamed Brisbane Athlon 64 X2 processors (from 90 nm SOI to 65 nm SOI), various generations of GPUs from both ATI and NVIDIA, and various generations of RAM and flash memory chips from Samsung, Toshiba and SK Hynix.

- In January 2010, Intel released Clarkdale Core i5 and Core i7 processors fabricated with a 32 nm process, down from a previous 45 nm process used in older iterations of the Nehalem processor microarchitecture. Intel, in particular, formerly focused on leveraging die shrinks to improve product performance at a regular cadence through its Tick-Tock model. In this business model, every new microarchitecture (tock) is followed by a die shrink (tick) to improve performance with the same microarchitecture.

- Die shrinks are beneficial to end-users as shrinking a die reduces the current used by each transistor switching on or off in semiconductor devices while maintaining the same clock frequency of a chip, making a product with less power consumption (and thus less heat production), increased clock rate headroom, and lower prices.

- Since the cost to fabricate a 200-mm or 300-mm silicon wafer is proportional to the number of fabrication steps, and not proportional to the number of chips on the wafer, die shrinks cram more chips onto each wafer, resulting in lowered manufacturing costs per chip.

**Half Shrink**

- In CPU fabrications, a die shrink always involves an advance to a lithographic node as defined by ITRS (shown in Table 3.1). For GPU and SoC manufacturing, the die shrink often involves shrinking the die on a node not defined by the ITRS, for instance the 150 nm, 110 nm, 80 nm, 55 nm, 40 nm and more currently 8 nm nodes, sometimes referred to as "half-nodes".

- This is a stopgap between two ITRS-defined lithographic nodes (thus called a "half-node shrink") before further shrink to the lower ITRS-defined nodes occurs, which helps save further R&D cost. The choice to perform die shrinks to either full-nodes or half-

nodes rests with the foundry and not the integrated circuit designer.

**Table 3.1**

| Half-shrink | |
| --- | --- |
| **Main ITRS node** | **Stopgap half-node** |
| 250 nm | 220 nm |
| 180 nm | 150 nm |
| 130 nm | 110 nm |
| 90 nm | 80 nm |
| 65 nm | 55 nm |
| 45 nm | 40 nm |
| 32 nm | 28 nm |
| 22 nm | 20 nm |
| 14 nm | 12 nm |
| 10 nm | 8 nm |
| 7 nm | 6 nm |
| 5 nm | 4 nm |
| 3 nm | N/A |

## 3.4 THE NANOSCALE MOSFET

- FETs have a few disadvantages like high drain resistance, moderate input impedance and slower operation. To overcome these disadvantages, the MOSFET which is an advanced FET is invented.

- MOSFET stands for Metal Oxide Silicon Field Effect Transistor or Metal Oxide Semiconductor Field Effect Transistor. This is also called as IGFET meaning Insulated Gate Field Effect Transistor. The FET is operated in both depletion and enhancement modes of operation. The following figure shows how a practical MOSFET looks like.

**Fig. 3.4**

**Construction of a MOSFET**

- The construction of a MOSFET is a bit similar to the FET. An oxide layer is deposited on the substrate to which the gate terminal is connected. This oxide layer acts as an insulator ($sio_2$ insulates from the substrate), and hence the MOSFET has another name as IGFET. In the construction of MOSFET, a lightly doped substrate, is diffused with a heavily doped region. Depending upon the substrate used, they are called as P-type and N-type MOSFETs.

- The following figure shows the construction of a MOSFET.

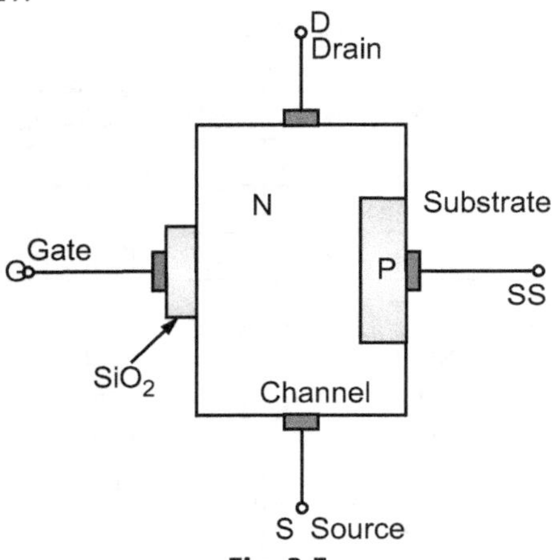


**Fig. 3.5**

- The voltage at gate controls the operation of the MOSFET. In this case, both positive and negative voltages can be applied on the gate as it is insulated from the channel. With negative gate bias voltage, it acts as depletion MOSFET while with positive gate bias voltage it acts as an Enhancement MOSFET.

**Classification of MOSFETs**

- Depending upon the type of materials used in the construction, and the type of operation, the MOSFETs are classified as in the following figure.

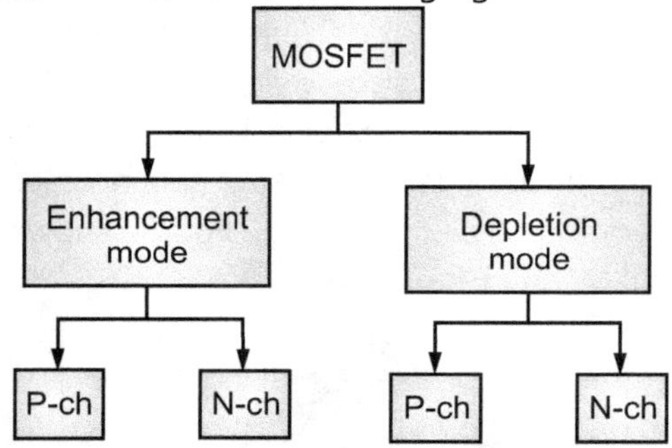


P-ch = P-channel    N-ch = N- channel

**Fig. 3.6**

- After the classification, let us go through the symbols of MOSFET.
- The N-channel MOSFETs are simply called as NMOS. The symbols for N-channel MOSFET are as given below.

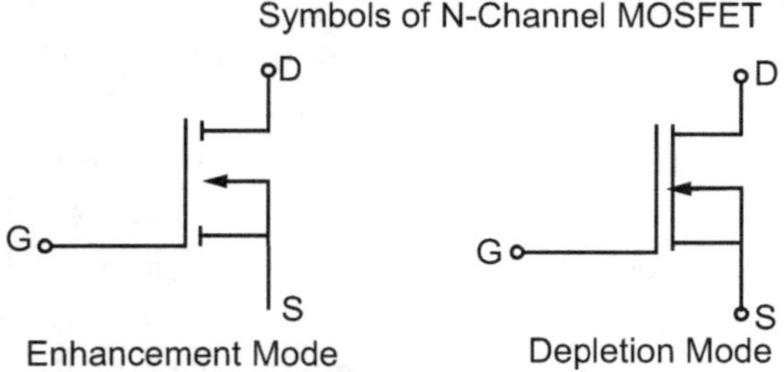


**Fig. 3.7**

- The P-channel MOSFETs are simply called as PMOS. The symbols for P-channel MOSFET are as given below.

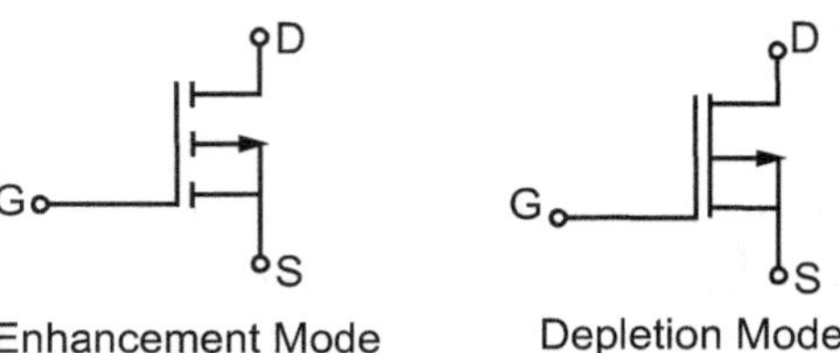


**Fig. 3.8**

- Now, let us go through the constructional details of an N-channel MOSFET. Usually an NChannel MOSFET is considered for explanation as this one is mostly used. Also, there is no need to mention that the study of one type explains the other too.

## Construction of N- Channel MOSFET

- Let us consider an N-channel MOSFET to understand its working. A lightly doped P-type substrate is taken into which two heavily doped N-type regions are diffused, which act as source and drain. Between these two N+ regions, there occurs diffusion to form an Nchannel, connecting drain and source.

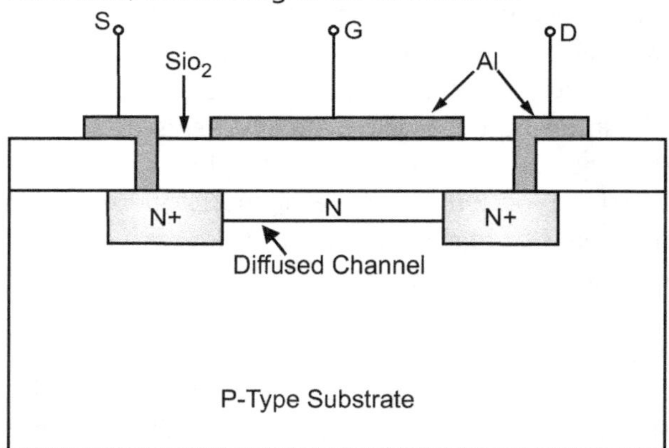


Structure of N-channel MOSFET

**Fig. 3.9**

- A thin layer of Silicon dioxide (SiO₂) is grown over the entire surface and holes are made to draw ohmic contacts for drain and source terminals. A conducting layer of aluminum is laid over the entire channel, upon this SiO₂ layer from source to drain which constitutes the gate. The SiO₂ substrate is connected to the common or ground terminals.
- Because of its construction, the MOSFET has a very less chip area than BJT, which is 5% of the occupancy when compared to bipolar junction transistor. This device can be operated in modes. They are depletion and enhancement modes. Let us try to get into the details.
- Working of N - Channel depletion mode depletion mode MOSFET

- For now, we have an idea that there is no PN junction present between gate and channel in this, unlike a FET. We can also observe that, the diffused channel N between two N+regions between two N+regions, the insulating dielectric $SiO_2$ and the aluminum metal layer of the gate together form a parallel plate capacitor.
- If the NMOS has to be worked in depletion mode, the gate terminal should be at negative potential while drain is at positive potential, as shown in Fig. 3.10.
- When no voltage is applied between gate and source, some current flows due to the voltage between drain and source. Let some negative voltage is applied at $V_{GG}$. Then the minority carriers i.e. holes, get attracted and settle near $SiO_2$ layer. But the majority carriers, i.e., electrons get repelled.

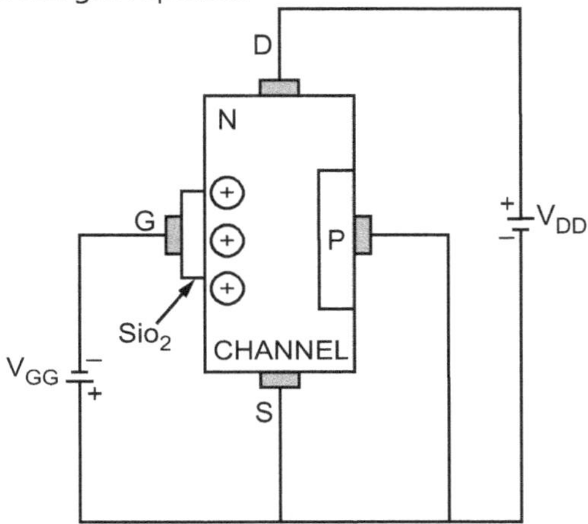


Working of MOSFET in depletion mode

**Fig. 3.10**

- With some amount of negative potential at $V_{GG}$ a certain amount of drain current $I_D$ flows through source to drain. When this negative potential is further increased, the electrons get depleted and the current $I_D$ decreases. Hence the more negative the applied $V_{GG}$, the lesser the value of drain current $I_D$ will be.
- The channel nearer to drain gets more depleted than at source likeinFETlikeinFET and the current flow decreases due to this effect. Hence it is called as depletion mode MOSFET.
- Working of N-Channel MOSFET Enhancement Mode Enhancement Mode
- The same MOSFET can be worked in enhancement mode, if we can change the polarities of the voltage $V_{GG}$. So, let us consider the MOSFET with gate source voltage $V_{GG}$ being positive as shown in the following Fig. 3.11.
- When no voltage is applied between gate and source, some current flows due to the voltage between drain and source. Let some positive voltage is applied at $V_{GG}$.

Then the minority carriers i.e. holes, get repelled and the majority carriers i.e. electrons gets attracted towards the $SiO_2$ layer.

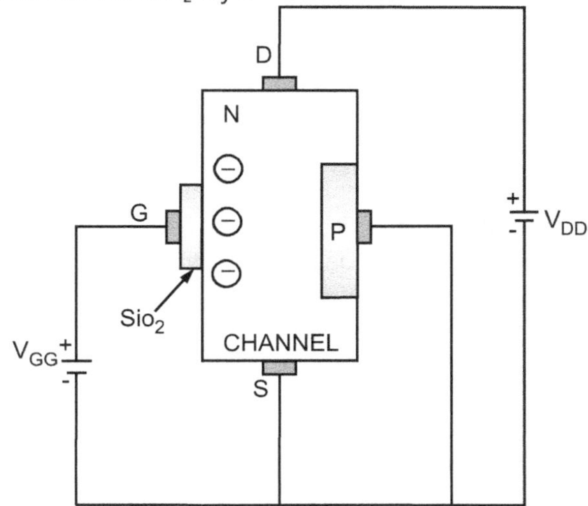


Working of MOSFET in Enhancement mode

**Fig. 3.11**

- With some amount of positive potential at $V_{GG}$ a certain amount of drain current $I_D$ flows through source to drain. When this positive potential is further increased, the current $I_D$ increases due to the flow of electrons from source and these are pushed further due to the voltage applied at $V_{GG}$. Hence the more positive the applied $V_{GG}$, the more the value of drain current $I_D$ will be. The current flow gets enhanced due to the increase in electron flow better than in depletion mode. Hence this mode is termed as Enhanced Mode MOSFET.

**P - Channel MOSFET**

- The construction and working of a PMOS is same as NMOS. A lightly doped n-substrate is taken into which two heavily doped P+ regions are diffused. These two P+ regions act as source and drain. A thin layer of $SiO_2$ is grown over the surface. Holes are cut through this layer to make contacts with P+ regions, as shown in the following Fig. 3.12.

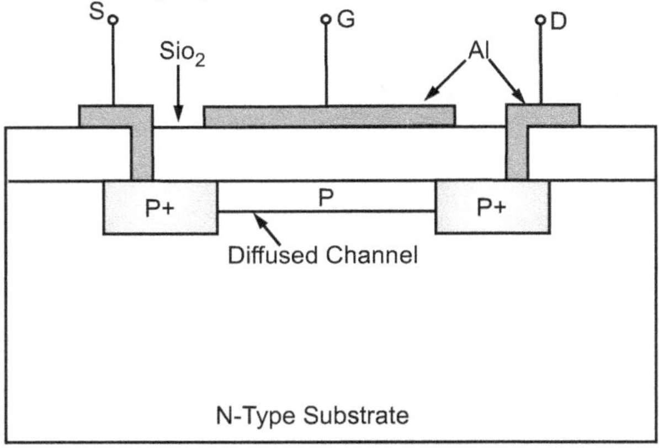


Structure of P-channel MOSFET

**Fig. 3.12**

## Working of PMOS

- When the gate terminal is given a negative potential at $V_{GG}$ than the drain source voltage $V_{DD}$, then due to the P+ regions present, the hole current is increased through the diffused P channel and the PMOS works in Enhancement Mode.
- When the gate terminal is given a positive potential at $V_{GG}$ than the drain source voltage $V_{DD}$, then due to the repulsion, the depletion occurs due to which the flow of current reduces. Thus PMOS works in Depletion Mode. Though the construction differs, the working is similar in both the type of MOSFETs. Hence with the change in voltage polarity both of the types can be used in both the modes.
- This can be better understood by having an idea on the drain characteristics curve.

## Drain Characteristics

- The drain characteristics of a MOSFET are drawn between the drain current $I_D$ and the drain source voltage $V_{DS}$. The characteristic curve is as shown below for different values of inputs.

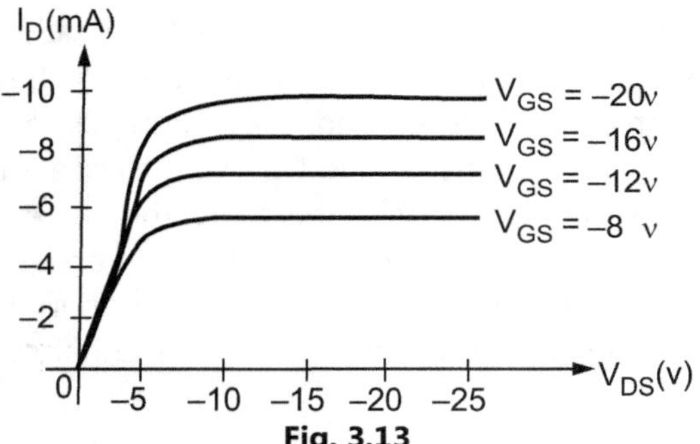


**Fig. 3.13**

- Actually when $V_{DS}$ is increased, the drain current $I_D$ should increase, but due to the applied $V_{GS}$, the drain current is controlled at certain level. Hence the gate current controls the output drain current.

## Transfer Characteristics

- Transfer characteristics define the change in the value of $V_{DS}$ with the change in $I_D$ and $V_{GS}$ in both depletion and enhancement modes. The below transfer characteristic curve is drawn for drain current versus gate to source voltage.

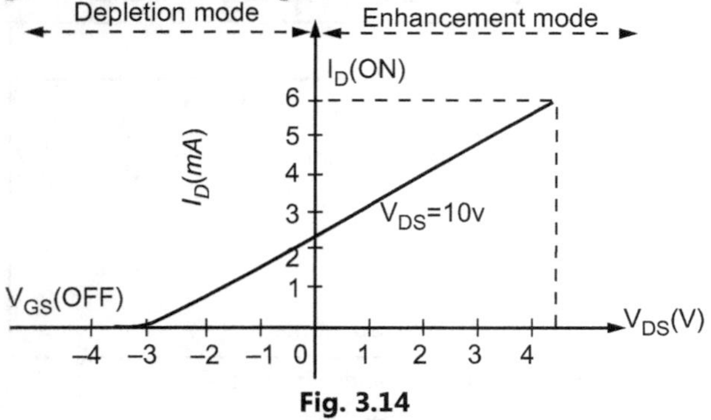


**Fig. 3.14**

### 3.4.1 Comparison between BJT, FET and MOSFET

Now that we have discussed all the above three, let us try to compare some of their properties.

| Terms | BJT | FET | MOSFET |
|---|---|---|---|
| Device type | Current controlled | Voltage controlled | Voltage Controlled |
| Current flow | Bipolar | Unipolar | Unipolar |
| Terminals | Not interchangeable | Interchangeable | Interchangeable |
| Operational modes | No modes | Depletion mode only | Both Enhancement and Depletion modes |
| Input impedance | Low | High | Very high |
| Output resistance | Moderate | Moderate | Low |
| Operational speed | Low | Moderate | High |
| Noise | High | Low | Low |
| Thermal stability | Low | Better | High |

So far, we have discussed various electronic components and their types along with their construction and working.

## 3.5 FINFETS

- The finFET is a transistor design, first developed by Chenming Hu and colleagues at the University of California at Berkeley, which attempts to overcome the worst types of short-channel effect encountered by deep submicron transistors, such as drain-induced barrer lowering (DIBL). These effects make it harder for the voltage on a gate electrode to deplete the channel underneath and stop the flow of carriers through the channel – in other words, to turn the transistor Off. By raising the channel above the surface of the wafer instead of creating the channel just below the surface, it is possible to wrap the gate around up to three of its sides, providing much greater electrostatic control over the carriers within it.
- There are a number of subtly different forms of trigate transistor structure that are being described as finFETs. The architecture typically takes advantage of self-aligned process steps to produce extremely narrow features that are much smaller than the wavelength of light generally used to pattern devices on a silicon

wafer. It is possible to create very thin fins - of 20 nm in width or less - on the surface of a silicon wafer using selective-etching processes, although they typically cannot currently be made less than 20 nm to 30nm because of the limits of lithographic resolution.

- The fin is used to form the raised channel. The gate is then deposited so that it wraps around the fin to form the trigate structure. As the channel is extremely thin the gate has much greater control over the carriers within it but, when the device is switched on, the shape limits the current through it to a low level. So, multiple fins are used in parallel to provide higher drive strengths.

- Originally, the finFET was developed for use on Silicon-On-Insulator (SOI) wafers. Recent developments have made it possible to produce working finFETs on bulk silicon wafers and improve the performance of certain parameters. The steep doping profile used to control leakage into the bulk substrate has a beneficial impact on DIBL, although increased doping has a negative impact on variability.

- Fully depleted SOI (*Guide*) transistors have been shown to offer comparable or better performance than finFETs. However, the relative compatibility of the bulk-silicon finFET with existing wafer fabrication processes and today's wafer-supply chain favors the finFET for high-volume IC production at 22 nm and below.

- FinFETs have key advantages over planar bulk devices. They exhibit more drive current per unit area than planar devices, largely because the height of the fin can be used to create a channel with a larger effective volume but still take advantage of a wraparound gate.

- The added performance capability of FinFETs can be used to achieve higher frequency numbers compared to bulk for a given power budget or lower power. The power reduces can come from two sources: reduces need for wide, high-drive standard cells; and the ability to operate with a lower supply voltage for a given amount of leakage.

## What Effect Does the FinFET have on Design?

- At such an early stage in its commercial development, the implications of the finFET are not entirely clear although results from Intel's work suggest that the impact on digital design need not be that great if conservative approaches are taken. At the International Solid State Circuits Conference (ISSCC) in 2012, Intel described its approach as being one largely of design migration from circuits created for planar processes, using modelling and simulation to assess how the transition from planar to trigate would affect circuit performance.

- A key difference between finFET-based design and that using conventional planar devices is that the freedom to choose the device's drive strength is reduced, especially for devices that are close to the minimum size. Drive strength can only be improved during layout by adding more fins. The effective width of the device becomes quantized, and the quantization effect is worse for smaller transistors for which the next step up from the minimum-size device is one that is twice as wide. In addition, the minimum number of fins may be two in practical manufacturing processes. This is due to the self-aligned spacer processes that are used to create fins at tight pitches – each sacrificial spacer element that is deposited creates a pair of fins.

- The Intel designers worked on the basis that whenever the optimum number of fins to achieve a particular drive strength was not an integer, they would round up to the next whole number – so that fractional fins were replaced with a full fin – rather than inserting transistors with less than optimal drive strength and risking the circuit not meeting timing.

- A team from Infineon Technologies and Texas Instruments reported at the International Solid State Circuits Conference (ISSCC) in 2006 that the problem of fin quantization was potentially a bigger issue for SRAMs – as they would only use one fin to save space – than in analog circuits, where the use of minimum-sized transistors is far less crucial. The problem of using minimum-sized devices throughout an SRAM can create problems for static noise margin – reducing the ability of the system to reliable read a memory cell. Ideally, the pass gates would be weaker than the pullup and pulldown devices, particularly the latter.

- One solution is to increase the fin count for pulldown devices but this increases area. Another is to weaken the pass gates by etching away the top surface of the gate – splitting the gate into a threshold-control and switching-control gate. This, however, increases process complexity. A third approach, as with sub-30nm planar CMOS processes is to use write-assist techniques – pushing the threshold voltage down temporarily by reducing the supply voltage.

- Designers working on experimental finFET processes have reported other problems, such as self-heating – a problem noted again by Infineon researchers, this time at the International Electron Device Meeting (IEDM) in

2009. Recent work by IBM, albeit on SOI wafers, has indicated that self-heating is not likely to be a major issue.

- FinFETs provide a number of advantages and several key disadvantages compared with bulk planar processes. Advantages include increased voltage headroom for circuits such as cascodes, lower gate resistance, which helps keep flicker noise under control, as well as improved matching, higher current drive and higher gain. However, the designer does not have the ability to control the channel as easily and the higher source/drain resistance cuts transconductance. On top of that, designers have little choice over voltages for I/O and have to develop more complex methods to achieve ESD immunity.

- A further impact on design is the need to consider layout density in circuits that are usually quite sparse compared with digital layouts. Device density variation can lead to dishing – similar to the problems of copper metallization encountered in the move to 130nm processes. Similarly, the fins at the edges of a cluster suffer higher variability than those in the middle. These effects lead to greater need for the use of dummy-fill shapes to reduce the variation in density. Foundry processes tend to put dummy fins at the end of each transistor stack.

- In terms of optimization for power, the finFET provides circuit designers with the opportunity to trade leakage for switching speed. Intel, for example, has deployed what it calls fast devices, with nominal leakage, medium-speed 'quarter-leakage' devices and slow 'tenth-leakage' devices. A problem facing process engineers is providing designers with a choice of threshold voltages to implement different circuits with different power-grade transistors at low cost.

- At the International Electron Devices Meeting in December 2012, Intel presented details of a family of finFET designsthat were optimized for high speed, low leakage and high (1.8V and 3.3V) operation in SoC designs.

- As the finFET was conceived to be a device with almost no channel doping and back biasing the gate is very area inefficient even where possible, the main technique for adjusting threshold is to manipulate the work function of the gate. An alternative that will push up variability and may have a knock-on effect on fin pitch – and therefore cell density – is to dope the channel.

- The lack of back or forward body bias control is one of the handicaps of today's finFET structures versus FD-SOI. However, the larger ecosystem for finFET-based designs has made it more difficult for FD-SOI to compete.

- Work is also underway at TSMC on introducing germanium into the fin of p-channel finFETs to improve the carrier mobility.

- The finFET may have other, more subtle effects on design, at least at the cell-library level and for analog designers. Design rules will be further restricted to allow gates and fins to be placed on a regular grid. A key issue is compatibility between fin pitch and the pitch of the intra and intercell routing layers, leading to non-integer heights for standard cells if counted in terms of M2 tracks.

- In their analysis of the finFET's influence on layout, Rob Aitken and colleagues and ARM found: "Fin and metal pitches have different scaling pressures, so they have not tended to line up. For example, at 14nm GlobalFoundries has stated that it uses a fin pitch of 48nm and a metal pitch of 64nm. The same values are used in TSMC's 16nm process."

- They added: "[Using GlobalFoundries' numbers, only certain integral combinations of fin and M2 pitch are possible: six, nine, twelve, etc. A 6 track cell height is unlikely to be viable for two reasons. First, it will contain at most four active fins (2N and 2P), and second there is unlikely to be enough room to route internal signals for complex cells such as flip-flops. A library containing 6 active fins (3N and 3P) would be ten fins tall. This equates to 10x48=480nm, which is equivalent to 7.5 M2 tracks. This is likely to be the smallest feasible library in this type of technology, and comes with the obvious issues relating to non- integral track heights for physical design."

- In their analysis of routing techniques for sub-28nm processes, CMU and IBM researchers performed simulations to look at analog designs on finFETs that revealed issues with restricted design rules. "Our design simulations based on preliminary models reveal that FinFETs have a mixed impact on analog circuits. The restricted design rules expected to be seen in sub 20nm nodes greatly limit the allowable channel lengths for analog designers, which in turn can be worked around by stacking devices in series to emulate a long channel transistor, however, at the expense of increased parasitic capacitance.

- "Conversely, FinFETs offer improved electrostatic control which translates to a higher intrinsic gain. Given these and several factors, some analog circuit topologies that have been considered obsolete may need to be revisited. For instance, topologies such as high gain linear amplifiers can be used in conjunction with switched capacitor circuits to balance performance for variation tolerance. Furthermore, emerging post-silicon tuning techniques such as self-healing and statistical element selection appear to be extremely valuable."

- Synopsys has a useful discussion on the practicalities of designing with finFETs, which is summarised here.

**When can we use finFETs?**

- Unless you work for Intel or a research group with access to customized processes, there is no way to implement finFET-based designs commercially. This is expected to change with the move to 14 nm processes, with the Common Platform foundry alliance (Global Foundries, IBM, Samsung) effectively committing to this shift in early 2012. Global Foundries has said it will introduce finFETs in its 14 nm process the devices will be optimized for mobile systems. The world's largest foundry, TSMC, has yet to say when it will introduce finFETs although the technology is likely to be in place for the 14 nm and may be brought forward to 20 nm.

**What are the risks of using finFETs?**

- The difficulties of dealing with a new technology 3D transistor design in terms of parasitic extraction and physical behavior, the major issue is cost : cost of software is very expensive building a finFET uses a number of additional steps in a manufacturing flow that is already struggling to contain the cost of advanced lithography : double patterning in the next few years, and possibly a move to EUV lithography in the second half of the decade. Figures presented by Qualcomm at IEDM 2013 indicated that the jump in cost to finFET was lower than that caused by the shift to double patterning from 28 nm to 20 nm. The Back-End of Line (BEOL) processes are more or less the same for 20, 16 nm and 14 nm technologies provided by the foundries.

## 3.5.1  FinFET : The Promises and the Challenges

- While the new multi-gate or tri-gate architectures, also known as FinFET technology, deliver superior levels of scalability, design engineers face significant challenges in creating designs that optimize the promise of this exciting new technology. Jamil Kawa, group director of the Solutions Group, Synopsys, and Andy Biddle, product marketing manager, Galaxy Implementation Platform, Synopsys, explain how Synopsys is working with foundry partners and design teams to help them accelerate innovation and get the best out of their investments in FinFETs.

- Design metrics including performance, power, area, cost and time to market have not changed since the inception of the Integrated Circuit (IC) industry. In fact, Moore's law is all about optimizing those parameters by driving to the smallest possible transistor size with each new technology generation. However, as process technologies continued to shrink towards 20-nanometers (nm), it became impossible to achieve a similar scaling of certain device parameters, the power supply voltage, which is the dominant factor in determining dynamic power. When in additional optimizing for one variable such as performance automatically translated to unwanted compromises in other areas like power.

- Given the new emerging metric of performance per unit power (Koomey's law), one major design optimization alternative designers have in FinFETs, as compared to planar technology, is much better performance at the same power budget, or equal performance at a much lower power budget.

- From Moore's Law, we can infer that FinFETs represent the most radical shift in semiconductor technology in over 40 years. When Gordon Moore came up with his "law" back in 1965, he had in mind a design of about 50 components. Today's chips consist of billions of transistors and design teams strive for "better, sooner, cheaper" products with every new process node. However, as feature sizes have become finer, the perils of high leakage current due to short-channel effects and varying dopant levels have threatened to derail the industry's progress to smaller geometries.

- The FinFET transistor structure promises to rejuvenate the chip industry by rescuing it from the short-channel effects that limit device scalability faced by current planar transistor structures.

## 3.5.2  FinFET : A Technology Primer

- FinFETs have their technology roots in the 1990s, when DARPA looked to fund research into possible successors to the planar transistor. A UC Berkeley team led by Dr. Chenming Hu proposed a new structure for the transistor that would reduce leakage current.

- The Berkeley team suggested that a thin-body MOSFET structure would control short-channel effects and suppress leakage by keeping the gate capacitance in closer proximity to the whole of the channel. They proposed two possible structures as shown in following Fig. 3.15.

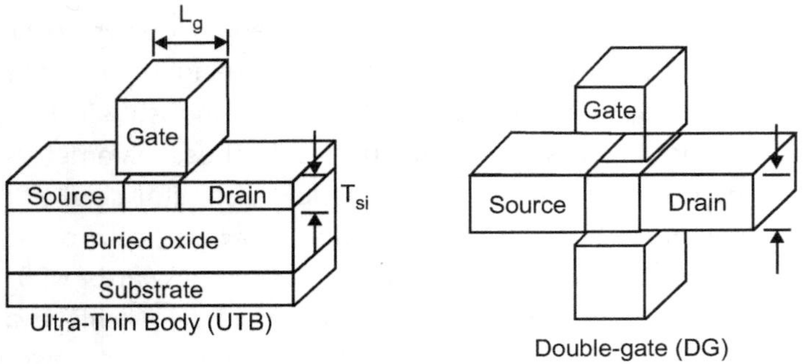


**Fig. 3.15 : Thin-body MOSFETs are the origin of today's FinFET transistors**

- Rotating the DG structure, which has the potential to provide the lowest gate leakage current, enables easier manufacturing using standard lithography techniques as the gate electrodes become self-aligned and the layout is similar to that of a planar FET as shown in Fig. 3.16.

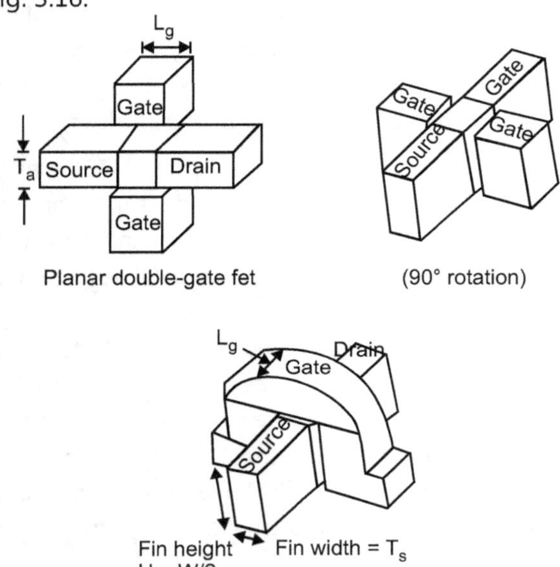


**Fig. 3.16 : From planar DG FET to FinFET**

- Modern FinFETs are 3D structures that rise above the planar substrate, giving them more volume than a planar gate for the same planar area. Planar area given to the excellent control of the conducting channel by the gate, which "wraps" around the channel, very little current is allowed to leak through the body when the device is in the off state. This allows the use of lower threshold voltages, which results in optimal switching speeds and power.

- Other research teams have shown that FinFETs are scalable as long as it is possible to scale the thickness of the channel. For example, KAIST has demonstrated a 3 nm FinFET in its lab.

**The FinFET Promise :**

- Leading foundries estimate the additional processing cost of 3D devices to be 2% to 5% higher than that of the corresponding Planar wafer fabrication. FinFETs are estimated to be up to 37% faster while using less than half the dynamic power or cut static leakage current by as much as 90%.

- FinFET also promise to problematic performance of the parameters such as power tradeoff etc. Designers can run the transistors faster and use the same amount of power, compared to the planar equivalent, or run them at the same performance using less power. This enables design teams to balance throughput, performance and power to match the needs of each application.

**Design Challenges or Minimizing Impact :**

- Increasingly, designers care less about packing more transistors on a die (Moore's Law) and more about delivering the best performance per Watt for the application. Interestingly, Jonathan Koomey has shown that the energy efficiency of computers has doubled nearly every 18 months since the first computers were built in the 1950s. Koomey's Law expands on Moore's law, especially given that quoting channel lengths is becoming less relevant. Given the abundance of transistors per unit area in advanced nodes, designers would use multi-processors at lower voltage to get the same throughput of a fewer number of processors at a higher voltage, sacrificing some additional area for the sake of saving power at the same throughput level.

- The foundries want to make the transition to FinFET processes as transparent and smooth as possible for the design community. In this case EDA and IP industries need to work behind the scenes to ensure that the tools understand and model the complexities involved. Design teams want to take advantage of the power, performance and area benefits that FinFETs offer while still getting to market quickly and painlessly through a familiar process of creating the RTL and taking it through a backend implementation process.

**IP Design Challenges : Not Just Another Transistor :**

- While developers can take a familiar design flow and work with FinFET technology in much the same way as their previous bulk CMOS designs, the quality of results they achieve will depend to a large extent on the quality of the IP they choose.

- Developing optimized memory and standard cells (physical IP) for FinFET requires expertise and experience. An experienced design team will be able to exploit the features that FinFET structures offer in order to create the best physical IP and not leave any power savings or performance on the table.

- In order to continue on the path of Moore's Law – and Koomey's Law – designers must be able to leverage the target technology for maximum benefit, and invariably, that means focusing on the details

- Synopsys has been working with industry and academic partners for several years to gain a detailed understanding of FinFET technology, and apply that knowledge to develop IP, tools and services for successful FinFET design.

- As a leading tool and IP developer, Synopsys is uniquely qualified to provide specific FinFET tool methodologies and FinFET-based memory and standard cell IP to customers developing differentiated leading-edge products in a broad range of applications from mobile computing to enterprise.

- As we move to FinFET, one of the challenges is the discrete size of the fin. Transistor width (W), which is one of the main variables for tweaking transistor sizes, is no longer a continuum. Discrete fin sizing brings a new variable in design, without any easy workarounds, that designers have never had to deal with before.

- Furthermore, additional design levers usually utilized by the IP designer, such as varying the channel length or body biasing, are either much more restrictive or are of limited benefit due to the intrinsic characteristics of FinFET technology.

- Another challenge has to do with the complexity of the model. The FinFET is a 3D structure that has a lot of subdivided resistance and capacitance compared with a planar structure. This 3D structure requires a more complex model and more data manipulation than planar transistors.

- The complexity of the model has implications for the whole backend flow including extraction, layout, DRC and LVS, for the engineers responsible for managing the design. Experience counts when it comes to optimizing FinFET designs efficiently in order to achieve the best quality of results.

**Experience Counts :**

- When it comes to IP design, getting the new FinFET technology requires experience. Synopsys has spent several years understanding the characteristics of FinFET technology and applying those to create new standard cell architectures and memory compilers. Synopsys has successfully navigated through complex FinFET issues. For example, there are specific challenges related to read-write access for memories. Synopsys has exploited the inherently low operating voltages of FinFETs to enable the design of memories with low retention voltages.

- Another fundamental issue that determines a transistor's performance is its stress profile – the mechanical stress that we deliberately introduce into the device to enhance its performance. Because of its vertical fin, the FinFET has a significantly different stress profile from a planar transistor. Synopsys has been collaborating with industry partners from an early stage to apply its Technology Computer-Aided Design (TCAD) tools to the task of accurately modeling FinFET stress profiles (for more information, see TCAD Tools).

- Synopsys continues to work closely with the major foundries to accurately capture all of the efficiency of FinFET technology, and to create models that we can use within the entire design flow from concept to implementation, including SPICE modeling, extraction and physical IP design.

### 3.5.3 The FinFET Tool Story

- The finFET tool is a transparent transition, it allows user to scale design to increasingly smaller geometry processes. This technology will require implementation of the tools to minimizing power consumption and maximize the clock speed and utilization.

- FinFETs require some specific enhancements made in the following areas :

  1. TCAD Tools.
  2. Mask Synthesis.
  3. Transistor Models.
  4. SPICE Simulation Tools.
  5. RC Extraction Tools.
  6. Physical Verification Tools.

**1. TCAD Tools :**

- To harness the full potential of 3D FETs, wafer processing technologies are being developed to controllably dope the fin sidewalls and stress the fins to boost device performance. To support these efforts, TCAD tools are used by the foundries during development to guide and optimize the semiconductor fabrication process.

- An important example of the need for 3D TCAD simulation is in the process optimization of SRAM cells, where stress and doping proximity effects require that all transistors comprising the SRAM be simulated in a single structure. This is made possible by recent advances in 3D structure generation, mesh generation and parallel algorithms.

- The small geometries targeted for FinFETs have introduced a concern with the impact of process variability on device and circuit performance. While these effects were negligible on higher geometry processes, they are now becoming first order effects. These variations caused by random dopant fluctuations, line edge roughness, layout induced stress, and other process variations ultimately manifest themselves as variations in device performance, in particular with threshold voltage shifts and local currents that impact timing and power. TCAD tools are used to simulate these effects used by EDA tools.

- Synopsys has deployed Sentaurus TCAD in FinFET research and development since 2005 at leading foundries, Integrated Device Manufacturers (IDMs) and research universities, and has made highly complex and sophisticated refinements to these tools as a result of this collaboration. These refinements include changes to our plasma-doping model, fin dimensional optimization to achieve device performance targets and modeling of the random process variations to improve device performance. Fig. 3.17 shows an example of the 3D simulation performed by Sentaurus for p-channel FinFETs.

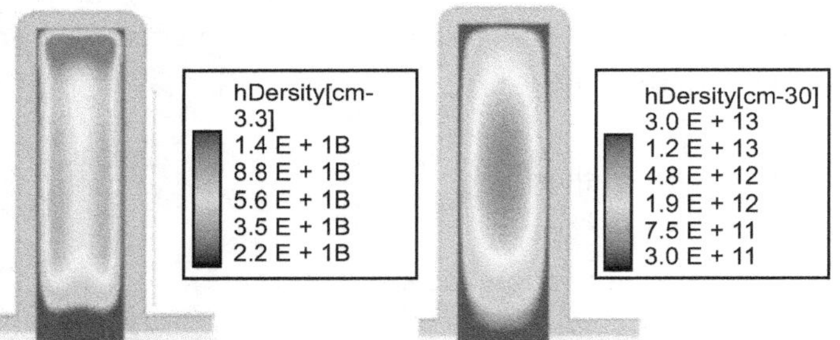


$I_{on}$ (left) and $I_{off}$ (right) in Fin cross section

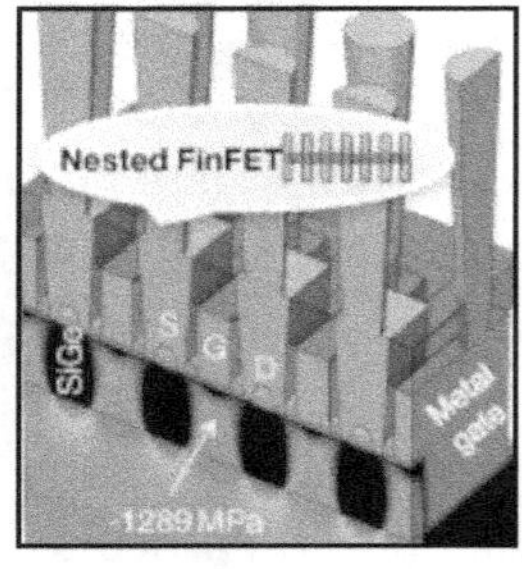


**Fig. 3.17 : Stress fields in pFinFETs simulated with TCAD Sentaurus**

## 2. Mask Synthesis :

- Mask synthesis is a key component in advanced manufacturing. This mask synthesis process is used to post- process the resultant layouts produced by EDA tools and effects in the lithography process used in manufacturing. The advanced geometries targeted for FinFET are expected to require Self-Aligned Double Patterning (SADP) in deposition manufacturing steps to create the fins rather than defining the fins lithographically. As the fins are tall and thin, traditional lithography/OPC methods would result in line-edge roughness problems.

- The Synopsys product provides a comprehensive and powerful environment for performing full-chip proximity correction, building models for correction and analyzing proximity effects on corrected and uncorrected IC layout patterns. These products are the mask synthesis tool of choice for IDMs and foundries building FinFET-based designs. Synopsys is closely engaged with the foundries on refining and deploying the Proteus SADP solution.

## 3. Transistor Models :

- FinFETs introduce much higher complexities for resistance and parasitic capacitance. Additional information is needed in the model for source/drain resistance extensions, contact resistances fringing effects and the wider number of coupling capacitances introduced by the three dimensional structures. The new behaviors are captured in new standardized models used by spice simulators.

- The Berkeley Short-channel IGFET Model for Common Multi-Gate (BSIM-CMG) compact model is used by SPICE simulators to ensure accurate simulation of designs using these new devices.

## 4. SPICE Simulation Tools :

- The Synopsys of SPICE and Fast SPICE simulators have been used extensively by the leading FinFET foundries to validate correct and accurate functionality with the new BSIM-CMG models. These tools form the corner stone for transistor level library and circuit design.

- The Synopsys HSPICE simulator has been selected as the gold standard for foundries introducing FinFETs. Synopsys has multiple simulators supporting the BSIM-CMG models, HSPICE and FineSim SPICE for full SPICE accuracy and CustomSim and FineSim Pro for Fast SPICE use.

## 5. Resistance/Capacitance (RC) Extraction Tools :

- The 3D nature of FinFETs and the multiple fins making up the transistors introduce a large number of new parasitic resistance and capacitances to be considered, modeled and extracted from the FinFET-based designs. Fig. 3.18 shows some of the parasitics introduced by this technology.

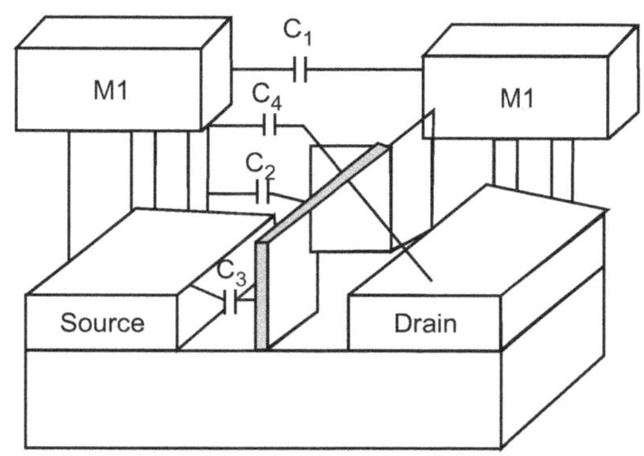

$C_1$ : S contact to D contact     $C_3$ : Gate to S/D diffusion
$C_2$ : Gate to S/D contact     $C_4$ : S contact to D diffusion

**Fig. 3.18 : FinFET Parasitics**

- The interconnect modeling of semiconductors has been standardized in the open source Interconnect Technology Format (ITF). This format has recently been extended to add the FinFET requirements.

- Synopsys' StarRC extraction tool has been enhanced to support the new ITF models and is extensively used in the extraction of FinFET-based designs. StarRC is certified by leading FinFET foundries and is the industry standard for signoff extraction.

**6. Physical Verification Tools :**

- Physical verification is another tools affected by FinFET technology. The new runsets used by the physical verification tools are used to verify Logic versus Schematic (LVS) correctness, and Design Rule Checks (DRCs).

- FinFETs require LVS enhancements to support recognition of these new devices in the layout and enable parameter extraction and identification of proximity effects. Other LVS enhancements include new source-drain resistance calculations. A number of new design rules have been introduced including fin-to-fin spacing and fin widths.

- Synopsy's IC Validator physical verification product has been enhanced to support LVS and DRC for FinFETs. It is currently being used for the development of FinFET-based designs and IP.

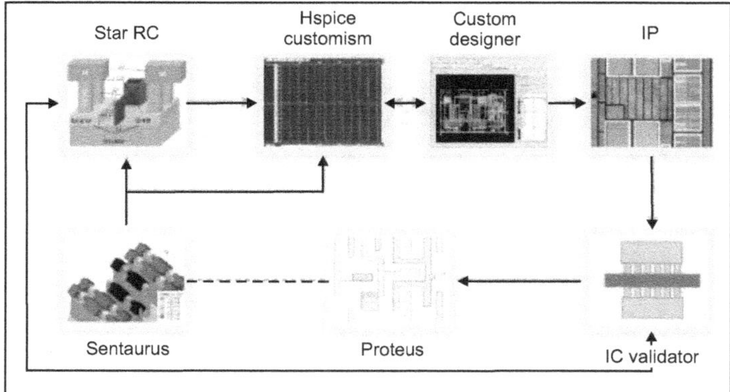


**Fig. 3.19 : Synopsys FinFET Technology**

- The Synopsys TCAD and complete Galaxy Implementation Platform of tools, including IC Compiler, Galaxy Custom Designer and PrimeTime, have already been used to tapeout 3D FET production designs and numerous test chips. Synopsys tools are ready for the next wave of FinFET technology adopters. Fig. 3.19 shows the entire design flow from concept to implementation, including SPICE modeling, extraction and physical IP design.

- Transition and significant benefits for circuit designer can be used finFET technology or synospys tools and IP. Synopsys is leading the industry in its efforts to create IP, tools, flows and expertise that will guide the design community towards the successful adoption of this radical shift in semiconductor technology.

- Historically, design teams have transitioned their IP from older planar technologies to the latest process nodes by using their in-house design capabilities and IP re-use. FinFET technology has created new challenges for many of these design teams because their current tools and techniques may not enable them to design their IP optimally for FinFET processes, delaying time to market. FinFETs require a new generation of design experience, expertise and tools in order to get the most from the technology.

- Synopsys has extensive experience and expertise with FinFETs and can help design teams to mitigate their risk in developing FinFET-based IP processes. As well as being an early developer of a vast portfolio of physical IP for FinFET, Synopsys is currently working alongside foundry partners and customer design teams to help them design highly differentiated products in order to win in highly competitive markets.

### 3.5.4 Construction of FinFET

- FinFET technology has been born as a result of the relentless increase in the levels of integration. According to Moore's law has held true for many years from the earliest years of integrated circuit technology. Essentially it states that the number of transistors on a given area of silicon doubles every two years.

- Some of the landmark chips of the relatively early integrated circuit era had a low transistor count even though they were advanced for the time. The 6800 microprocessor for example had just 5000 transistors. Todays have many orders of magnitude more.

- To achieve the large increases in levels of integration, many parameters have changed. Fundamentally the feature sizes have reduced to enable more devices to be fabricated within a given area. However other figures such as power dissipation, and line voltage have reduced along with increased frequency performance.

- There are limits to the scalability of the individual devices and as process technologies continued to shrink towards 20 nm, it became impossible to achieve the proper scaling of various device parameters. Those like the power supply voltage, which is the dominant factor in determining dynamic power were particularly affected. It was found that optimising for one variable such as performance resulted in unwanted compromises in other areas like power. It was therefore necessary to look at other more revolutionary options like a change in transistor structure from the traditional planar transistor.

### 3.5.5 Properties of FinFET

- FinFET technology takes its name from the fact that the FET structure used looks like a set of fins when viewed.

- The main characteristic of the FinFET is that it has a conducting channel wrapped by a thin silicon "fin" from which it gains its name. The thickness of the fin determines the effective channel length of the device.

- In terms of its structure, it typically has a vertical fin on a substrate which runs between a larger drain and source area. This protrudes vertically above the substrate as a fin.

- The gate orientation is at right angles to the vertical fin. And to traverse from one side of the fin to the other it wraps over the fin, enabling it to interface with three side of the fin or channel.

- This form of gate structure provides improved electrical control over the channel conduction and it helps reduce leakage current levels and overcomes some other short-channel effects.

- The term FinFET is used somewhat generically. Sometimes it is used to describe any fin-based, multigate transistor architecture regardless of number of gates.

### 3.5.6 Advantages of FinFET Technology

There are several advantages to IC manufacturers of using FinFETs.

- **Power :** Much lower power consumption allows high integration levels. Early adopters reported 150% improvements.

- **Operating Voltage :** FinFETs operate at a lower voltage as a result of their lower threshold voltage.

- **Feature Sizes :** Possible to pass through the 20nm barrier previously thought as an end point.

- **Static Leakage Current :** Typically reduced by up to 90%

- **Operating Speed :** Often in excess of 30% faster than the non-FinFET versions.

- The FinFET is a technology that is used within ICs. FinFETs are not available as discrete devices. However FinFET technology is becoming more widespread as feature sizes within integrated circuits fall and there is a growing need to provide very much higher levels of integration with less power consumption within integrated circuits.

## 3.6 VERTICAL MOSFETS

- The VMOS transistor, named after the V-shaped groove, is a vertical MOSFET with high current handling capability as well as high blocking voltage. It consists of a double diffused n+/p layer, which is cut by a V-shaped groove as shown in Fig. 3.20 (a). The V-groove is easily fabricated by anisotropically etching a (100) silicon surface using a concentrated KOH solution.

- The V-groove is then coated with a gate oxide, followed by the gate electrode. As the V-groove cuts through the double diffused layer, it creates two vertical MOSFETs, one on each side of the groove. The combination of the V-groove with the double diffused layers results in a short gatelength, which is determined by the thickness of the p-type layer. The vertical structure allows the use of a low-doped drain region, which results in a high blocking voltage.

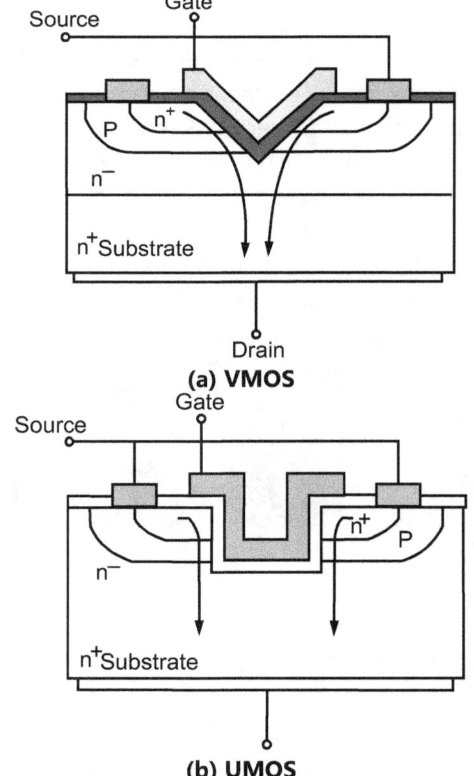


**Fig. 3.20 : Cross-section of two vertical MOSFET structures**

- Another alternate structure is the UMOS structure. A vertical trench is etched though the double diffused layer, again resulting in two vertical MOSFETs.
- Either one of these vertical structures can further be combined with the HEXFET layout. This layout resembles a honeycomb structure in which the hexagonal areas are source areas, while the gate metal is located on the perimeters.

## 3.7 LIMITS TO SCALING

- Over the past three decades, CMOS technology scaling has been a primary driver of the electronics industry and has provided a path toward both denser and faster integration. The transistors manufactured today are 20 times faster and occupy less than 1% of the area of those built 20 years ago.
- The number of devices per chip and the system performance has been improving exponentially over the last two decades. As the channel length is reduced, the performance improves, the power per switching event decreases, and the density improves. But the power density, total circuits per chip, and the total chip power consumption has been increasing.
- The need for more performance and integration has accelerated the scaling trends in almost every device parameter, such as lithography, effective channel length, gate dielectric thickness, supply voltage, device leakage, etc. Some of these parameters are approaching fundamental limits, and alternatives to the existing material and structures may need to be identified in order to continue scaling.

In scaling there are really two issues

1. Devices Can we Build Smaller Devices
2. What will their performance be try to avoid the wet noodle effect Wires

There is concern about our ability to scale both of these Components.

### 1. Limitations

Limitations to device scaling has been around since working in 3m nMOS, 22 years ago (actually bipolar)

- Worries were
  - ➢ Short channel effect
  - ➢ Punchthrough
- Drain control of current rather than gate
  - ➢ Hot electrons
  - ➢ Parasitic resistances

Now worries are a little different

- Oxide tunnel currents

- Punchthrough
- Parameter control
- Parasitic resistances

### 2. Transistor Scaling

- People are building very short channel devices
- Shown are I-V curves for 15nm L pMOS
- And a short channel nMOS
- The structure is strange
- FinFET
- But you can make them work

### 3. Wire Scaling

- More uncertainty than transistor scaling
- Many options with complex trade-offs
- For each metal layer. Need to set H, TT, TB, e1, e2, conductivity of the metal

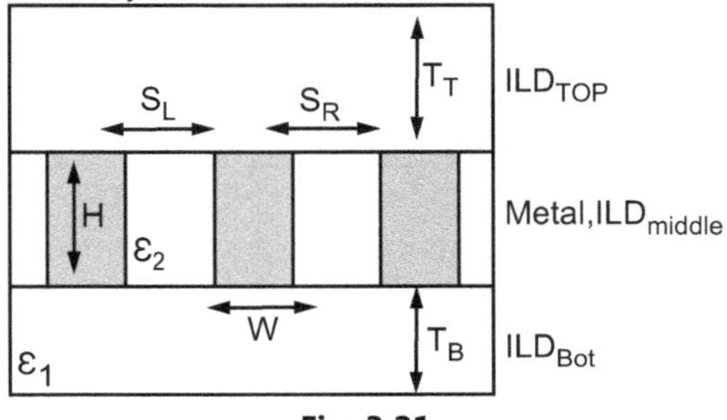


**Fig. 3.21**

## 3.8 SYSTEM INTEGRATION LIMITS

As microprocessors become more complex due to technology scaling, microprocessor designers have encountered several challenges which force them to think beyond the design plane, and look ahead to post-silicon:

- **Process Variation :** As photolithography techniques get closer to the fundamental laws of optics, achieving high accuracy in doping concentrations and etched wires is becoming more difficult and prone to errors due to variation. Designers now must simulate across multiple fabrication process corners before a chip is certified ready for production, or use system-level techniques for dealing with effects of variation.
- **Stricter Design Rules :** Due to lithography and etch issues with scaling, design rules for layout have become increasingly stringent. Designers must keep in mind an ever increasing list of rules when laying out custom circuits. The overhead for custom design is now reaching a tipping point, with many design houses opting to switch to electronic design automation (EDA) tools to automate their design process.
- **Timing / Design Closure :** As clock frequencies tend to scale up, designers are finding it more difficult to

distribute and maintain low clock skew between these high frequency clocks across the entire chip. This has led to a rising interest in multicore and multiprocessor architectures, since an overall speedup can be obtained even with lower clock frequency by using the computational power of all the cores.

- **First-Pass Success :** As die sizes shrink (due to scaling), and wafer sizes go up (due to lower manufacturing costs), the number of dies per wafer increases, and the complexity of making suitable photomasks goes up rapidly. A mask set for a modern technology can cost several million dollars. This non-recurring expense deters the old iterative philosophy involving several "spin-cycles" to find errors in silicon, and encourages first-pass silicon success. Several design philosophies have been developed to aid this new design flow, including design for manufacturing (DFM), design for test (DFT), and Design for X.

## 3.9 INTERCONNECT ISSUES

- It's well known that advanced chips contain billions of transistors – this is an incredible, mind-blowing fact to be sure – but did you know that large-scale integrated chips (about the size of a fingernail) can contain ~30 miles of interconnect "wires" in stacked levels? These wires function like highways or pipelines to transport electrons, connect transistors and other components to each other, and make them functional.

- And just as the speed you can drive your sports car depends (at least in part) on how clogged the freeway is, chip performance depends on the ability to move signals and power through these ultra-tiny wires. In fact, as the shrinking of feature dimensions (scaling) has continued, interconnects are now becoming the speed bottleneck in today's most advanced chips. So, let's spend a little time learning about them.

### Interconnect Layers

- During the first portion of chip-making (or front-end-of-line), the individual components (transistors, capacitors, etc.) are fabricated on the wafer. In the back-end-of-line, these components are connected to each other to distribute signals, as well as power and ground. There simply isn't room on the chip surface to create all those connections in a single layer, so chip manufacturers build vertical levels of interconnects. While simpler integrated circuits (ICs) may have just a few metal layers, complex ICs can have ten or more layers of wiring.

- Interconnects close to the transistors need to be small, as they attach/join to the components that are

themselves very small and often closely packed together. These lower-level lines – called local interconnects – are usually thin and short in length. Global interconnects are higher up in the structure; they travel between different blocks of the circuit and are thus typically thick, long, and widely separated. Connections between interconnect levels, called vias, allow signals and power to be transmitted from one layer to the next.

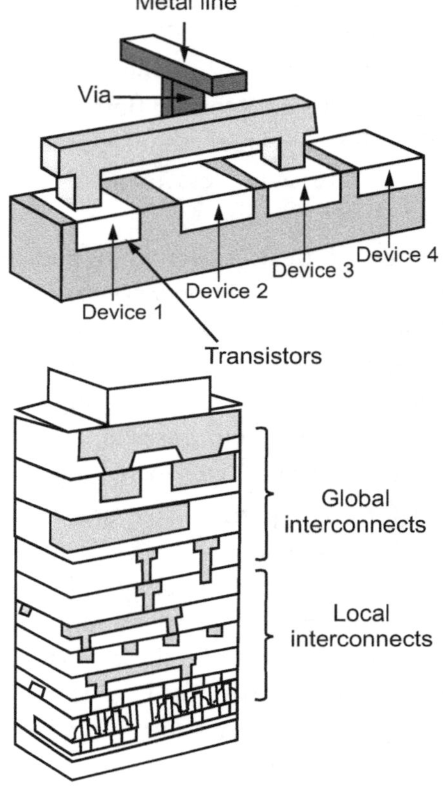


**Fig. 3.22**

### Interconnect Materials

- For decades, aluminum interconnects were the industry standard. To create these interconnects, a layer of aluminum was deposited. Then, the metal was patterned and etched, and insulating material was deposited to separate the conducting lines. In the late 1990s, chipmakers switched to copper, which conducts electricity better than aluminum – somewhat similar to how a copper-bottomed frying pan heats up faster than an all-aluminum pan.

- Higher conductivity lines improved overall IC performance. In addition, copper lines could be made smaller, keeping pace with transistor size scaling. Copper wiring is also more durable and reliable. However, creating copper interconnects is much more complex, and a whole new manufacturing scheme had to be developed for this technology inflection.

- The copper interconnect process starts with deposition of an insulating (dielectric) material, for example silicon dioxide, followed by the creation of trenches. The trenches are then filled with copper (using chemical/electroplating technologies), and the excess is removed to form a flat surface for subsequent processing.

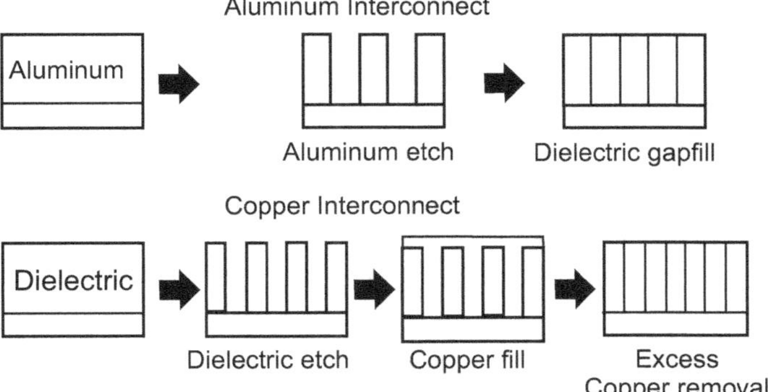


**Fig. 3.23**

## The RC Challenge

- Over the years, transistors have decreased dramatically in size. As transistors have gotten smaller and smaller, interconnects also have had to scale in size. Today, we're at the point where conventional copper interconnects are facing a significant roadblock to further scaling, and that roadblock is known as the RC challenge. Let's pause to define these two variables.

- The electrical resistance (R) of a material describes how difficult it is to move electrical current through a particular cross-section of that material, which is a function of the orientation and proximity of the material's atoms. Capacitance (C) refers to a material's ability to store electrical charge. The product of resistance and capacitance (RC) needs to be low to create fast chips since device speed is inversely proportional to RC (lower RC = faster devices).

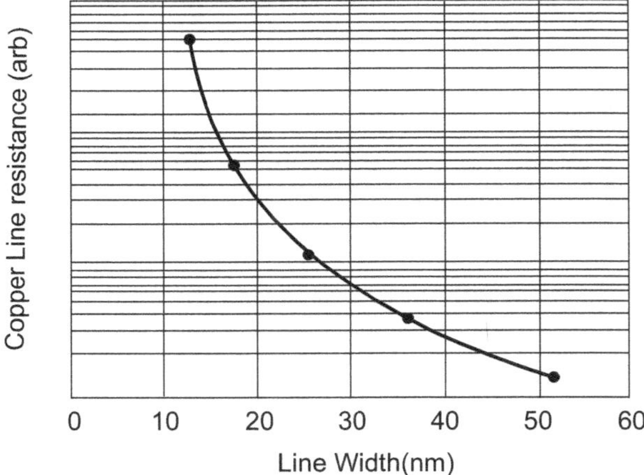


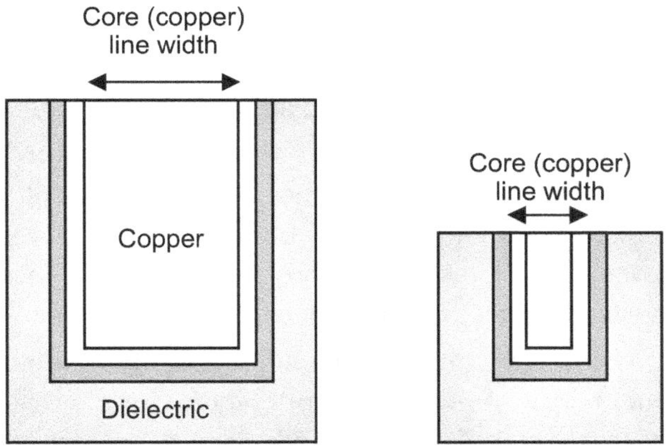


**Fig. 3.24**

- Looking at the "R" side of this challenge, higher-resistance lines carry less current, which slows device speed. This is because higher resistance reduces electron flow, so it takes longer to build up the minimum charge (number of electrons) or "threshold voltage" at a transistor's "gate" to turn it on. While transistor speed continues to improve with scaling (reducing the distance electrons must travel), the challenge for interconnect scaling is to not become a bottleneck and lose that performance improvement by slowing the flow of electrons between transistors.

- On the "C" side, capacitance is a function of the insulating dielectric material around the metal lines and the distance between them. Higher capacitance slows electrons and can create unwanted "cross talk," where the signal (voltage change) in one metal line influences the signal in a neighboring line and causes the device to malfunction.

- In addition to maintaining a suitable distance between lines, the development of "low-k" dielectric materials (capacitance is a function of a material's "k value") has significantly lowered capacitance. Today's dielectric materials average around k = 2.5, compared with k = 4.2 for pure silicon dioxide. Various methods exist to achieve lower values; however, the resulting ultra-low-k films become increasingly fragile as the k value decreases, posing additional challenges for use in manufacturing.

- Scaling Solutions and Future Directions To address these issues for further scaling, the industry continues to identify ways to manage RC, particularly focusing on the metals involved. Copper has been successfully used for multiple device generations, so the industry is investing significant effort in developing new approaches to extend its use.

- Creating copper lines involves a series of layers — typically a tantalum nitride barrier (prevents metal diffusion into the dielectric), a tantalum liner (improves barrier adherence to the metal), a copper seed layer (initiates the metal fill/plating), and finally, the bulk (core conducting) copper metal.

- A key area of development focus is identifying strategies for improving the barrier/liner/seed layers to lower overall resistance and enable smaller lines by creating "space" for the bulk copper fill.

- One strategy is to make the high-resistance barrier and liner thinner. However, opportunity for further thinning of these layers is limited. Both need to be continuous (no gaps or voids in the film) in order to achieve good reliability performance. This requires a minimum thickness of about 1.5 nm to 2 nm for each layer, which leads to a combined thickness of 3-4 nm on both sides of the trench structures.

- A potential alternative being studied is a new type of "self-forming" barrier that reacts with and forms on the dielectric surface adjacent to the copper line, which allows more room for copper. Also, new liners made of cobalt and ruthenium are being developed to replace tantalum.

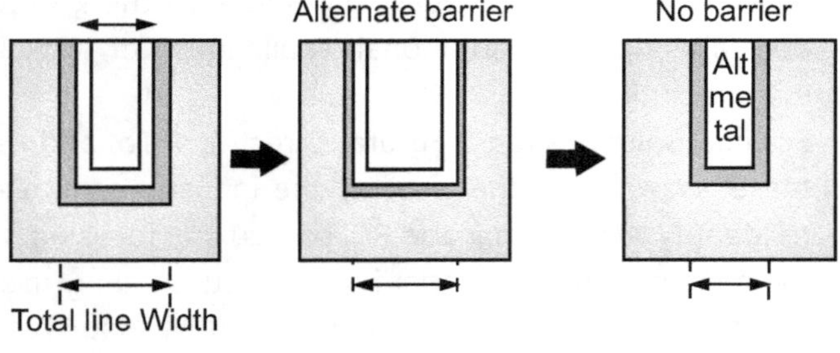


**Fig. 3.25**

- They adhere better to the copper seed, enabling it to be more conformal (eliminating voids) and thinner. Already, new technologies are in place to achieve void-free copper fill of small trenches. Around the 5 nm technology node, however, copper as the primary conducting metal will ultimately have to be replaced with a conducting material that does not require a barrier for these ultra-thin lines.

- While much attention is placed on the metals, there's also some investigation into improvements on the dielectric side. Here, the holy grail is to reduce the dielectric constant as much as possible, the ultimate being k = 1, which is air. Indeed, novel developments using "air gaps" have been used, but both fabrication and production cost challenges are significant.

- Thus, many of the ideas being explored for interconnect scaling involve the development of new metals, designs, and processes – and those new technologies that are in the "pipeline" are sure to make ever smaller, faster connections a reality.

## SUMMARY

- The term shrink (sometimes optical shrink or process shrink) refers to the scaling of metal-oxide-semiconductor (MOS) devices. The act of shrinking a die is to create a somewhat identical circuit using a more advanced fabrication process, usually involving an advance of lithographic nodes.

- This reduces overall costs for a chip company, as the absence of major architectural changes to the processor lowers research and development costs, while at the same time allowing more processor dies to be manufactured on the same piece of silicon wafer, resulting in less cost per product sold.

## EXERCISE

1. Explain Shrink-down approaches.
2. Describe CMOS Scaling.
3. Explain the nanoscale MOSFET.
4. Draw and explain Finfets.
5. Draw and explain operation of Vertical MOSFETs.
6. What are the limits of scaling?
7. Explain system integration limits.
8. Describe interconnect issues.

# NANO ELECTRONICS SEMICONDUCTOR DEVICES

## 4.1 INTRODUCTION

- Nano electronics refers to the use of nanotechnology in electronic components. The term covers a diverse set of devices and materials, with the common characteristic that they are so small that inter-atomic interactions and quantum mechanical properties need to be studied extensively. Some of these candidates include: hybrid molecular / semiconductor electronics, one-dimensional nanotubes / nanowires (e.g. Silicon nanowires or Carbon nanotubes) or advanced molecular electronics.

- Nano electronic devices have critical dimensions with a size range between 1 nm and 100 nm. Recent silicon MOSFET (metal-oxide-semiconductor field-effect transistor, or MOS transistor) technology generations are already within this regime, including 22 nanometer CMOS (complementary MOS) nodes and succeeding 14 nm, 10 nm and 7 nm FinFET (fin field-effect transistor) generations. Nanoelectronics are sometimes considered as disruptive technology because present candidates are significantly different from traditional transistors.

- In 1965 Gordon Moore observed that silicon transistors were undergoing a continual process of scaling downward, an observation which was later codified as Moore's law. Since his observation, transistor minimum feature sizes have decreased from 10 micrometers to the 7 nm range (as of 2019). The field of nanoelectronics aims to enable the continued realization of this law by using new methods and materials to build electronic devices with feature sizes on the nanoscale.

**Approaches**

### 1. Nanofabrication

- For example, electron transistors, which involve transistor operation based on a single electron. Nano electromechanical systems also fall under this category. Nanofabrication can be used to construct ultradense parallel arrays of nanowires, as an alternative to synthesizing nanowires individually. Of particular prominence in this field, Silicon nanowires are being increasingly studied towards diverse applications in Nano electronics, energy conversion and storage. Such SiNWs can be fabricated by thermal oxidation in large quantities to yield nanowires with controllable thickness.

### 2. Nanomaterials Electronics

- Besides being small and allowing more transistors to be packed into a single chip, the uniform and symmetrical structure of nanowires and/or nanotubes allows a higher electron mobility (faster electron movement in the material), a higher dielectric constant (faster frequency), and a symmetrical electron/hole characteristic.

- Also, nanoparticles can be used as quantum dots.

### 3. Molecular Electronics

- Single molecule devices are another possibility. These schemes would make heavy use of molecular self-assembly, designing the device components to construct a larger structure or even a complete system on their own. This can be very useful for reconfigurable computing, and may even completely replace present FPGA technology.

- Molecular electronics is a new technology which is still in its infancy, but also brings hope for truly atomic scale electronic systems in the future. One of the more promising applications of molecular electronics was proposed by the IBM researcher Ari Aviram and the theoretical chemist Mark Ratner in their 1974 and 1988 papers *Molecules for Memory, Logic and Amplification,* (see Unimolecular rectifier).

- This is one of many possible ways in which a molecular level diode / transistor might be synthesized by organic chemistry. A model system was proposed with a spiro carbon structure giving a molecular diode about half a nanometre across which could be connected by polythiophene molecular wires. Theoretical calculations showed the design to be sound in principle and there is still hope that such a system can be made to work.

**Other Approaches**

- Nanoionics studies the transport of ions rather than electrons in nanoscale systems.

- Nanophotonics studies the behavior of light on the nanoscale, and has the goal of developing devices that take advantage of this behavior.

## 4.2 RESONANT TUNNELING DIODE

- A Resonant-Tunneling Diode (**RTD**) is a diode with a resonant-tunneling structure in which electrons can tunnel through some resonant states at certain energy levels. The current–voltage characteristic often exhibits negative differential resistance regions.

- All types of tunneling diodes make use of quantum mechanical tunneling. Characteristic to the current–voltage relationship of a tunneling diode is the presence of one or more negative differential resistance regions, which enables many unique applications. Tunneling diodes can be very compact and are also capable of ultra-high-speed operation because the quantum tunneling effect through the very thin layers is a very fast process. One area of active research is directed toward building oscillators and switching devices that can operate at terahertz frequencies.

- An RTD can be fabricated using many different types of materials (such as III–V, type IV, II–VI semiconductor) and different types of resonant tunneling structures, such as the heavily doped p–n junction in Esaki diodes, double barrier, triple barrier, quantum well, or quantum wire. The structure and fabrication process of Si/SiGe resonant interband tunneling diodes are suitable for integration with modern Si complementary metal–oxide–semiconductor (CMOS) and Si/SiGe heterojunction bipolar technology.

- One type of RTDs is formed as a single quantum well structure surrounded by very thin layer barriers. This structure is called a double barrier structure. Carriers such as electrons and holes can only have discrete energy values inside the quantum well. When a voltage is placed across an RTD, a terahertz wave is emitted, which is why the energy value inside the quantum well is equal to that of the emitter side. As voltage is increased, the terahertz wave dies out because the energy value in the quantum well is outside the emitter side energy.

- Another feature seen in RTD structures is the negative resistance on application of bias as can be seen in the image generated from Nanohub. The forming of negative resistance will be examined in detail in operation section below.

- This structure can be grown by molecular beam heteroepitaxy. GaAs and AlAs in particular are used to form this structure. AlAs/InGaAs or InAlAs/InGaAs can be used.

- The operation of electronic circuits containing RTDs can be described by a Liénard system of equations, which are a generalization of the Van der Pol oscillator equation.

**Operation of RTD :**

- The following process is also illustrated from rightside figure. Depending on the number of barriers and number of confined states inside the well, the process described below could be repeated.

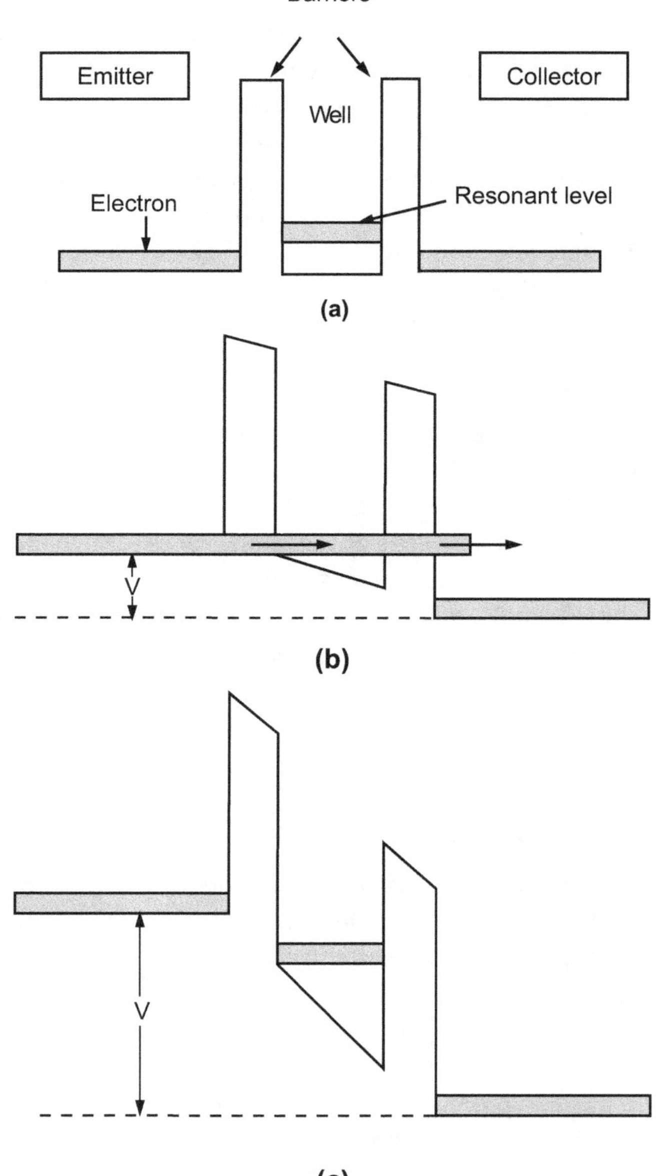

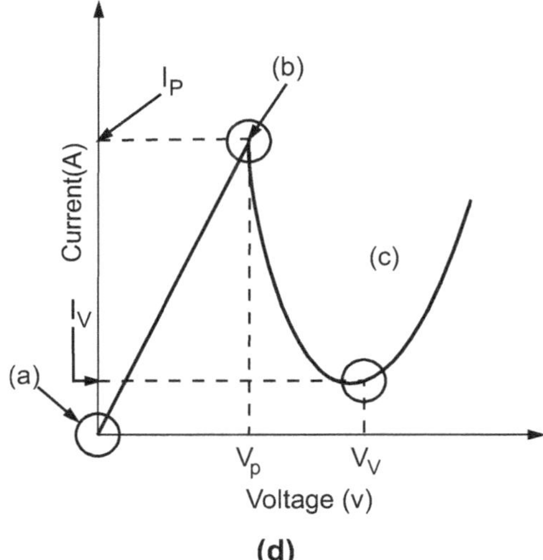

**(d)**

**Fig. 4.1 : The band strucure of the RTD under different applied bias voltages**

## Positive Resistance Region

- For low bias, as the bias increase, 1st confined state between the potential barriers is getting closer to the source Fermi level, so the current it carries increases.

## Negative Resistance Region

- As bias increases further, the 1st confined state becomes lower in energy and gradually goes into the energy range of bandgap, so the current it carries decreases. At this time, the 2nd confined state is still too high above in energy to conduct significant current.

## Second Positive Resistance Region

- Similar to the first region, as the 2nd confined state becomes closer and closer to source Fermi level, it carries more current, causing the total current to increase again.

- The band strucure of the RTD under different applied bias voltages is shown in Fig. 4.2. The device consists of two tunnel barriers enclosing a quantum well. Outside the barriers, doped contacts form a Fermi sea of electrons. Inside the well there is a resonant level of some small width. Under low bias voltage (Fig. 4.2 (a)), a small current flows due to non-resonant and scattering assisted tunnelling, leakage current through surface states and thermionic emission over the barriers.

- When the bias voltage is increased, the emitter level rises relative to the resonant level. Peak current (Fig. 4.2 (b)) is achieved when the emission region conduction band is at the same energy as the resonant level and the most electrons tunnel from the emission region into the resonant level.

- If the voltage is increased further, the resonant level falls below the emission region (Fig. 4.2 (c)) and the current is reduced, leading to the NDR property observed in Fig. 4.1. Eventually background effects dominate, and the current rises again as in a traditional resistor.

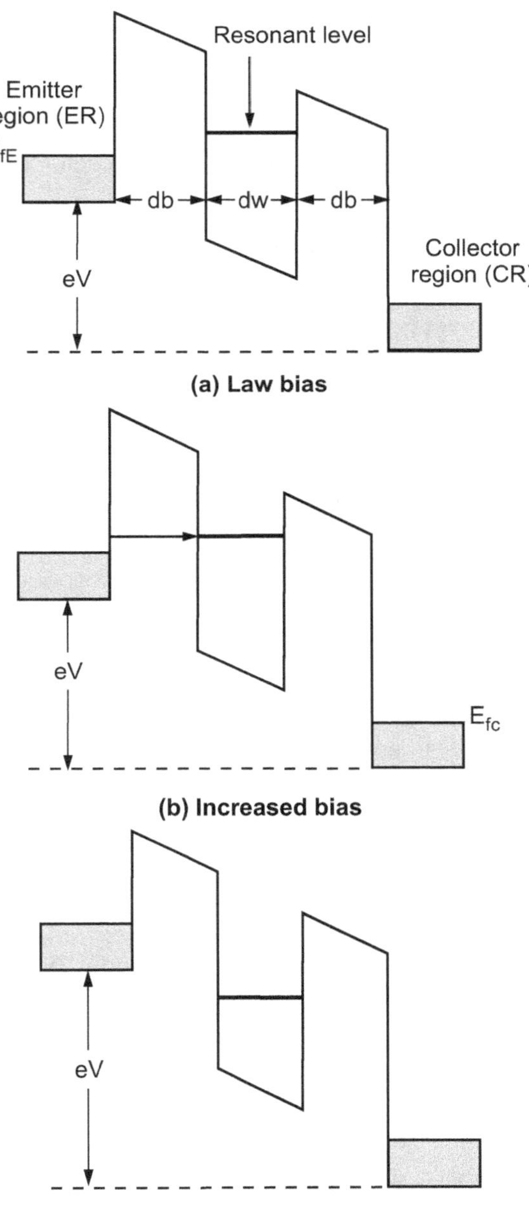


**(a) Law bias**

**(b) Increased bias**

**(c) Further increased bias**

**Fig. 4.2 : Band structure of the RTD under different applied bias voltages**

- Ultrahigh-speed circuit applications of resonant-tunneling diodes (RTDs) have been developed. The key points are the utilization of the edge-triggered and latching properties arising from the negative differential resistance of the RTD, and the combination of RTDs and high electron mobility transistors (HEMTs).

High-speed and low power operation of various flip-flop (FF) circuits monolithically integrating.

- InP-based RTDs and HEMTs have been demonstrated at room temperature, including a delayed flip-flop operation at 35 Gbit/s. By extending the concept of electronic-input circuits to the optical-input circuit, an ultrahigh-speed optoelectronic circuit integrating RTDs and a photodiode has been developed. Using this optoelectronic circuit, we have succeeded in demultiplexing a 80 Gbit/s optical signal into a 40 Gbit/s electrical signal. These results show the potentiality of RTD-based circuits for ultrahigh-speed communications and signal processing circuits.

## 4.3 QUANTUM DOTS

- A Quantum dot is a nanostructure or tiny particle of a semiconducting substance with a diameter range of 2nm to 10nm. They were initially discovered in 1980 & rsquos. These nano products display exclusive electronic properties, intermediate amid those of discrete molecules and bulk semiconductors. Its electronic features are determined by the shape and size.

- These semiconductor materials can be manufactured from an element, such as germanium or silicon or also from compound such as CdSe or CdS. There are many quantum dots supplier you can find in the market, however you need to be careful while purchasing from them.

- Quantum Dots (QDs) are tiny semiconductor particles a few nanometres in size, having optical and electronic properties that differ from larger particles due to quantum mechanics. They are a central topic in nanotechnology.

- When the quantum dots are illuminated by UV light, an electron in the quantum dot can be excited to a state of higher energy. In the case of a semiconducting quantum dot, this process corresponds to the transition of an electron from the valence band to the conductance band.

- The excited electron can drop back into the valence band releasing its energy by the emission of light. This light emission (photoluminescence) is illustrated in the figure on the right. The color of that light depends on the energy difference between the conductance band and the valence band.

- In the language of materials science, nanoscale semiconductor materials tightly confine either electrons or electron holes. Quantum dots are sometimes referred to as artificial atoms, emphasizing their singularity, having bound, discrete electronic states, like naturally occurring atoms or molecules.

- Quantum dots have properties intermediate between bulk semiconductors and discrete atoms or molecules. Their optoelectronic properties change as a function of both size and shape. Larger QDs of 5–6 nm diameter emit longer wavelengths, with colors such as orange or red. Smaller QDs (2–3 nm) emit shorter wavelengths, yielding colors like blue and green. However, the specific colors vary depending on the exact composition of the QD.

- Potential applications of quantum dots include single-electron transistors, solar cells, LEDs, lasers, single-photon sources, second-harmonic generation, quantum computing and medical imaging. Their small size allows for some QDs to be suspended in solution, which may lead to use in inkjet printing and spin-coating. They have been used in Langmuir-Blodgett thin-films. These processing techniques result in less expensive and less time-consuming methods of semiconductor fabrication.

- Quantum Dots (QDs) are man-made nanoscale crystals that that can transport electrons. When UV light hits these semiconducting nanoparticles, they can emit light of various colors. These artificial semiconductor nanoparticles that have found applications in composites, solar cells and fluorescent biological labels.

- Nanoparticles of semiconductors – quantum dots – were theorized in the 1970s and initially created in the early 1980s. If semiconductor particles are made small enough, quantum effects come into play, which limit the energies at which electrons and holes (the absence of an electron) can exist in the particles. As energy is related to wavelength (or color), this means that the optical properties of the particle can be finely tuned depending on its size. Thus, particles can be made to emit or absorb specific wavelengths (colors) of light, merely by controlling their size.

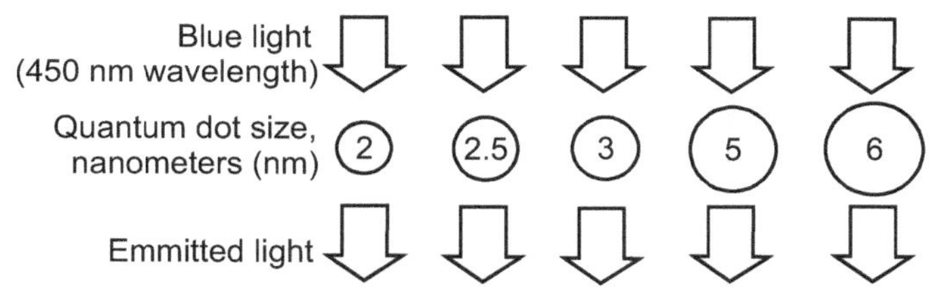


**Fig. 4.3**

- Quantum dots are nanoscale man-made crystals that have the ability to convert a spectrum of light into different colors. Each dot emits a different color depending on its size.

- Quantum dots are artificial nanostructures that can possess many varied properties, depending on their material and shape. For instance, due to their particular electronic properties they can be used as active materials in single-electron transistors.

- The properties of a quantum dot are not only determined by its size but also by its shape, composition, and structure, for instance if it's solid or hollow. A reliable manufacturing technology that makes use of quantum dots' properties – for a wide-ranging number of applications in such areas as catalysis, electronics, photonics, information storage, imaging, medicine, or sensing – needs to be capable of churning out large quantities of nanocrystals where each batch is produced according to the exactly same parameters.

- Because certain biological molecules are capable of molecular recognition and self-assembly, nanocrystals could also become an important building block for self-assembled functional nanodevices.

- The atom-like energy states of QDs furthermore contribute to special optical properties, such as a particle-size dependent wavelength of fluorescence; an effect which is used in fabricating optical probes for biological and medical imaging.

- So far, the use in bioanalytics and biolabeling has found the widest range of applications for colloidal QDs. Though the first generation of quantum dots already pointed out their potential, it took a lot of effort to improve basic properties, in particular colloidal stability in salt-containing solution. Initially, quantum dots have been used in very artificial environments, and these particles would have simply precipitated in 'real' samples, such as blood. These problems have been solved and QDs have found numerous use in real applications.

## 4.3.1 Quantum Dots in Medicine

- Quantum dots enable researchers to study cell processes at the level of a single molecule and may significantly improve the diagnosis and treatment of diseases such as cancers. QDs are either used as active sensor elements in high-resolution cellular imaging, where the fluorescence properties of the quantum dots are changed upon reaction with the analyte, or in passive label probes where selective receptor molecules such as antibodies have been conjugated to the surface of the dots.

- Quantum dots could revolutionize medicine. Unfortunately, most of them are toxic. Ironically, the existence of heavy metals in QDs such as cadmium, a well-established human toxicant and carcinogen, poses potential dangers especially for future medical application, where qdots are deliberately injected into the body.

- As the use of nanomaterials for biomedical applications is increasing, environmental pollution and toxicity have to be addressed, and the development of a non-toxic and biocompatible nanomaterial is becoming an important issue.

## 4.3.2 Quantum Dots in Photovoltaics

- The attractiveness of using quantum dots for making solar cells lies in several advantages over other approaches: They can be manufactured in an energy-saving room-temperature process; they can be made from abundant, inexpensive materials that do not require extensive purification, as silicon does; and they can be applied to a variety of inexpensive and even flexible substrate materials, such as lightweight plastics.

- Although using quantum dots as the basis for solar cells is not a new idea, attempts to make photovoltaic devices have not yet achieved sufficiently high efficiency in converting sunlight to power.

- A promising route for quantum dot solar cells is a semiconductor ink with the goal of enabling the coating of large areas of solar cell substrates in a single deposition step and thereby eliminating tens of deposition steps necessary with the previous layer-by-layer method.

### 4.3.3 Graphene Quantum Dots

- Graphene, which basically is an unrolled, planar form of a carbon nanotube therefore has become an extremely interesting candidate material for nanoscale electronics. Researchers have shown that it is possible to carve out nanoscale transistors from a single graphene crystal (i.e. graphene quantum dots). Unlike all other known materials, graphene remains highly stable and conductive even when it is cut into devices one nanometer wide.

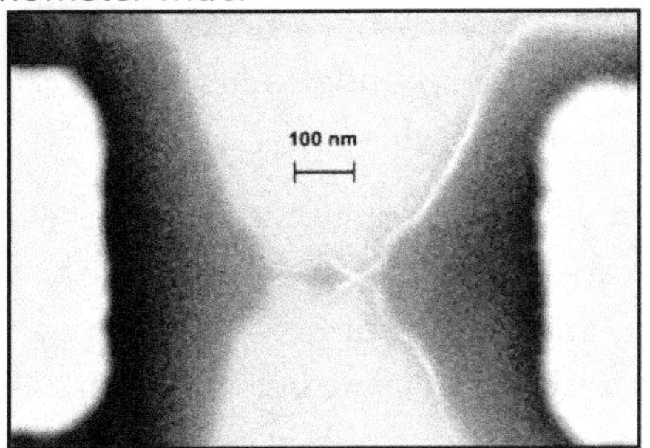


**Fig. 4.4 : Quantum dot carved from a graphene sheet.**
**(Image: Mesoscopic Physics Group, University of Manchester)**

- Graphene quantum dots (GQDs) also show great potential in the fields of photoelectronics, photovoltaics, biosensing, and bioimaging owing to their unique photoluminescence (PL) properties, including excellent biocompatibility, low toxicity, and high stability against photobleaching and photoblinking.
- Scientists still are working on finding efficient and universal methods for the synthesis of GQDs with high stability, controllable surface properties, and tunable PL emission wavelength.

### 4.3.4 Perovskite Quantum Dots

- Luminescent Quantum Dots (LQDs), which possess high photoluminescence quantum yields, flexible emission color controlling, and solution processibility, are promising for applications in lighting systems (warm white light without UV and infrared irradiation) and high quality displays.
- However, the commercialization of LQDs has been held back by the prohibitively high cost of their production. Currently, LQDs are prepared by the HI method, requiring at high temperature and tedious surface treating in order to improve both optical properties and stability.
- Although developed only recently, inorganic halide perovskite quantum dot systems have exhibited comparable and even better performances than traditional QDs in many fields.
- By preparing highly emissive inorganic perovskite quantum dots (IPQDs) at room temperature, IPQDs'

superior optical merits could lead to promising applications in lighting and displays.

### 4.3.5 Quantum Dot TVs and Displays

- The most commonly known use of quantum dots nowadays may be TV screens. Samsung and LG launched their QLED TVs in 2015, and a few other companies followed not long after.
- Quantum dots, because they are both photo-active (photoluminescent) and electro-active (electroluminescent) and have unique physical properties, will be at the core of next-generation displays. Compared to organic luminescent materials used in organic light emitting diodes (OLEDs), QD-based materials have purer colors, longer lifetime, lower manufacturing cost, and lower power consumption. Another key advantage is that, because QDs can be deposited on virtually any substrate, you can expect printable and flexible even rollable quantum dot displays of all sizes.

## 4.4 QUANTUM BLOCKADE

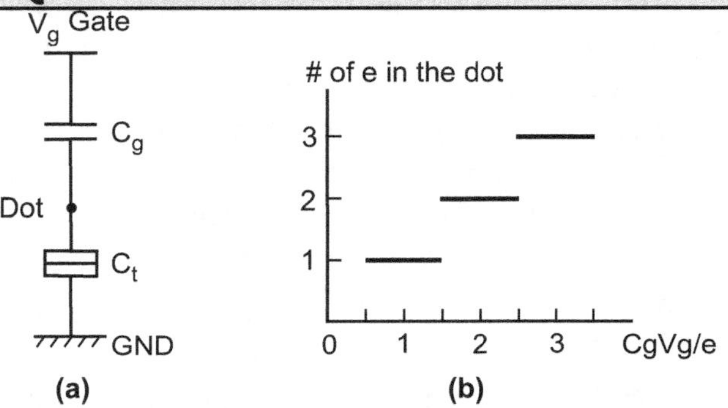


**Fig. 4.5 : (a) Equivalent circuit diagram of a single-electron box. (b) The number of the electron in the dot as a function of charge induced by the gate ($C_gV_g$).**

- The Coulomb blockade effect is the most fundamental phenomenon used not only in SETs/SHTs but also in single-electron memories, single-electron transport devices, and so on, to control the motion of a single electron. In this section, an electron system is mainly considered for simplicity, since the charge polarity is just the opposite in the hole case.
- Here, let us consider a small conductive dot connected to ground (large reservoirs as source) via tunnel junction as the simplest single-electron device, single-electron box, shown in Fig. 4.5 (a). Electrons can tunnel to the dot by applying voltage to the gate electrode which is capacitively coupled to the dot. If one electron tunnels to the dot, the increase of the electrostatic (Coulomb charging) energy $\Delta U$ in the system is expressed as

$$\Delta U = \frac{e^2}{2C_{dot}}$$

where $e$ is the elementary charge and $C_{dot}$ is the total capacitance of the dot. If the electric flux from the dot is all terminated at gate and ground, $C_{dot}$ is the sum of gate capacitance $C_g$ and tunnel junction capacitance $C_t$. When the dot becomes sufficiently small ($C_{dot}$ is small) and $\Delta U$ starts to exceed the thermal energy, even a single electron cannot tunnel to the dot without the help of external gate bias to overcome the Coulomb repulsion of the dot. This effect is called Coulomb blockade and it is the basic of the operation of SETs. When the Coulomb blockade works in the system, electrons are added to the dot one by one as the gate bias increases by $e/C_g$ in a step-like manner at zero temperature as shown in Fig. 4.5 (b). Note that, at finite temperature, this increase of the electron number is not a step function since it is blurred by the thermal energy.

## 4.5 SINGLE ELECTRON TRANSISTORS

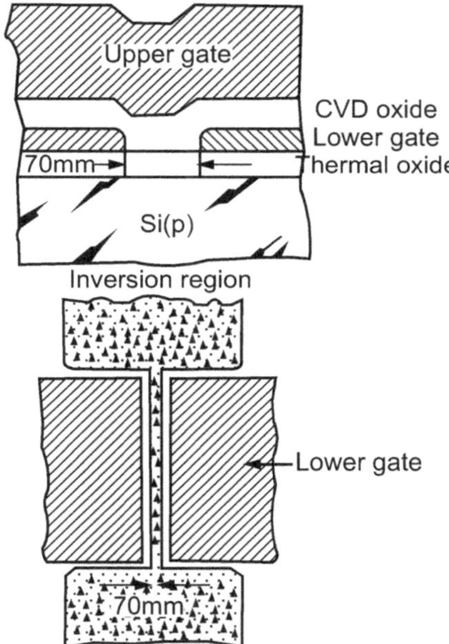


**Fig. 4.6 : Schematic diagram of the (a) cross-section and (b) top view of the silicon transistor with a continuous upper gate and a gap in the lower gate present**

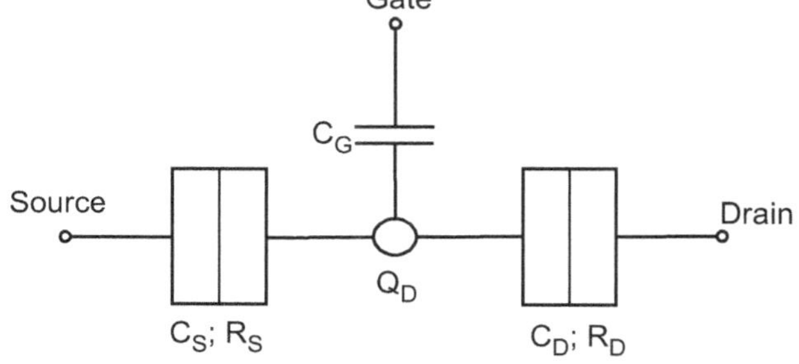


**Fig. 4.7 : Schematic of a basic SET and its internal electrical components**

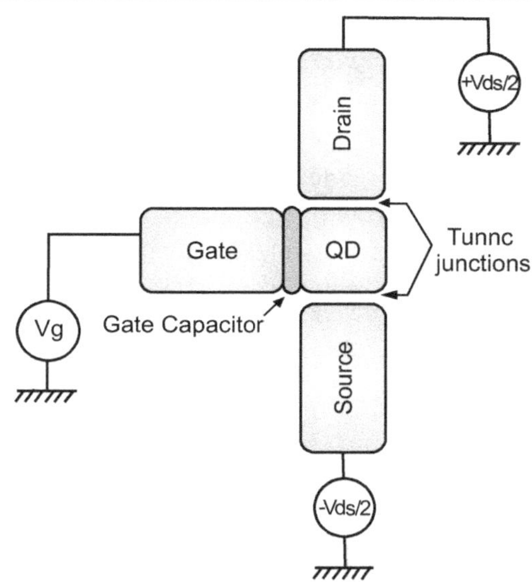


**Fig. 4.8 : Schematic of a single-electron transistor**

- Single-electron transistors (SETs) and single-hole transistors (SHTs) have attracted much attention for future integrated devices. SETs and SHTs have new functionalities that conventional complementary metal-oxide-semiconductor devices do not possess. We pay special attention to the high functionalities in silicon SET and SHT that operate at room temperature and are fabricated by very large scale integration compatible process. In this chapter, basic characteristics and fabrication process of room-temperature-operating SET and SHT are introduced. Furthermore, compact modeling of transport properties and additional functionalities are described for future-possible applications of SET and SHT for integrated circuits.

- Electronic and optical devices now employ sophisticated epitaxial growth techniques. Fine structures with nanometer accuracy are used in devices such as high-electron-mobility transistors, single electron transistors, high-efficiency laser diodes, and high performance light-emitting diodes. To achieve the required device performances, growth needs to be controlled very accurately. For this purpose, the crystal growth mechanisms must be understood on an atomistic scale.

- For microstructure fabrication, molecular beam epitaxy (MBE) and metal-organic vapor phase epitaxy (MOVPE) have been most frequently employed. MBE and MOVPE have different advantages for industrial use. MOVPE is sometimes more suitable for mass production, whereas MBE is more useful for devices with sophisticated and complicated structures. For understanding atomistic growth mechanisms, MBE is much better than MOVPE. In an MBE chamber, one can install various in situ monitoring systems, such as reflection high-energy electron diffraction (RHEED), a high-resolution scanning electron microscope (SEM), a reflection electron microscope (REM), and a mass spectroscopic analyzer (MS).

- MBE has other advantages for studying growth processes. The growth is conducted in a high vacuum, under which it is possible to get rid of the possible contaminating elements from the growth environments. Also, it is possible to use very pure materials that are commercially available for the growth elements. Ga metal of 7–9 s purity is not difficult to get. It is well known that even small amounts of impurities may strongly affect the growth; for instance, impurity atoms often help heterogeneous nucleation. A third advantage is that it is possible to realize very slow growth rates, such as one monolayer per second. This is possible because the level of the vacuum is so high that the growth rate can be slowed down to such a rate without oxygen contamination, which disturbs the normal epitaxial growth.

- Another advantage is that the gaseous species involved in the growth reaction is very simple. For instance, in the MBE of GaAs, the gaseous species are Ga vapor and $As_2$/$As_4$ molecules. This simplicity helps one to understand the atomistic growth mechanisms of MBE. With this knowledge, one can then understand the growth mechanisms of MOVPE and other growth methods, such as halogen vapor phase epitaxy—at least to some extent.

- Low-dimensional microstructures, such as quantum wires and quantum dots, have been attracting strong interest for their potential applications, such as high-performance semiconductor lasers and single-electron devices. Extensive studies have been carried out toward the fabrication of such devices. Epitaxial growth is one of the most hopeful techniques for fabricating the low-dimensional structures in semiconductors because it results in highly perfect nanostructures with high density. So far, many techniques of epitaxial growth have been proposed for the fabrication of nanostructures, such as epitaxial growth on patterned substrate, selective area epitaxy, and self-assembling growth. Among these techniques, epitaxial growth on the patterned substrate is one of the most promising techniques because one can control both the position and the size of the nanostructures.

- To control the shape of the microstructures, it is important to understand the elementary growth processes on the atomic scale. Surface diffusion between two surfaces (intersurface diffusion) is one of the most important processes that control the appearance and disappearance of a certain face. If the

intersurface diffusion occurs from the A facet to the B facet, the B facet grows faster and will disappear when these facets consist of a polyhedron. As an example of growing microstructure, the growth of a pyramid is studied by a high resolution scanning electron microscope installed in MBE. Real-time observation shows that the top of a GaAs pyramid changes from flat to spire (and vice versa) depending on the arsenic pressure. Growth elimination on a certain surface and enhancement of lateral growth, both by intersurface diffusion, are also present.

## 4.6 CARBON NANOTUBE ELECTRONICS

- Carbon Nanotubes (CNTs) are tubes made of carbon with diameters typically measured in nanometers. Carbon nanotubes often refers to single-wall carbon nanotubes (SWCNTs) with diameters in the range of a nanometer. They were discovered independently by Iijima and Ichihashi and Bethune et al. In carbon arc chambers similar to those used to produce fullerenes. Single-wall carbon nanotubes are one of the allotropes of carbon, intermediate between fullerene cages and flat graphene.

- Although not made this way, single-wall carbon nanotubes can be thought of as cutouts from a two-dimensional hexagonal lattice of carbon atoms rolled up along one of the Brava is lattice vectors of the hexagonal lattice to form a hollow cylinder. In this construction, periodic boundary conditions are imposed over the length of this roll up vector to yield a lattice with helical symmetry of seamlessly bonded carbon atoms on the cylinder surface.

- Carbon nanotubes also often refer to Multi-Wall carbon nanotubes (MWCNTs) consisting of nested single-wall carbon nanotubes. If not identical, these tubes are very similar to Oberlin, Endo and Koyama's long straight and parallel carbon layers cylindrically rolled around a hollow tube. Multi-wall carbon nanotubes are also sometimes used to refer to double- and triple-wall carbon nanotubes.

- Carbon nanotubes can also refer to tubes with an undetermined carbon-wall structure and diameters less than 100 nanometers. Such tubes were discovered by Radushkevich and Lukyanovich. While nanotubes of other compositions exist, most research has been focused on the carbon ones. Therefore, the "carbon" qualifier is often left implicit in the acronyms, and the names are abbreviated NT, SWNT, and MWNT.

- Carbon nanotubes can exhibit remarkable electrical conductivity. They also have exceptional tensile strength and thermal conductivity, because of their nanostructure and strength of the bonds between carbon atoms. In addition, they can be chemically modified. These properties are expected to be valuable in many areas of technology, such as electronics, optics, composite materials (replacing or complementing carbon fibers), nanotechnology, and other applications of materials science.

- Individual carbon nanotubes naturally align themselves into "ropes" held together by relatively weak van der Waals forces. The length of a carbon nanotube produced by common production methods is often not reported, but is much larger than its diameter. Although rare, nanotubes half a meter long have been created, with a length-to-diameter ratio of more than 100,000,000:1. For many purposes, the length of carbon nanotubes can be assumed to be infinite.

- Rolling up a hexagonal lattice along different directions to form different single-wall carbon nanotubes shows that all of these tubes have helical and translational symmetry along the tube axis and many also have nontrivial rotational symmetry about this axis. In addition, most are chiral, meaning the tube and its mirror image cannot be superimposed. This construction also allows single-wall carbon nanotubes to be labeled by a pair of small integers.

- A special group of achiral single-wall carbon nanotubes are metallic, but all the rest are either small or moderate band gap semiconductors. These electrical properties, however, do not depend of whether the tube is rolled up above or below the graphene plane and hence are the same for a tube and its mirror image.

## The Zigzag and Armchair Configurations

- In the study of nanotubes, one defines a "zigzag" path on a graphene-like lattice as a path that turns 60 degrees, alternating left and right, after stepping through each bond. It is also conventional to define an "armchair" path as one that makes two left turns of 60 degrees followed by two right turns every four steps.

- On some carbon nanotubes, there is a closed zigzag path that goes around the tube. One says that the tube is of the zigzag type or configuration, or simply is a zigzag nanotube. If the tube is instead encircled by a closed armchair path, it is said to be of the armchair type, or an armchair nanotube.

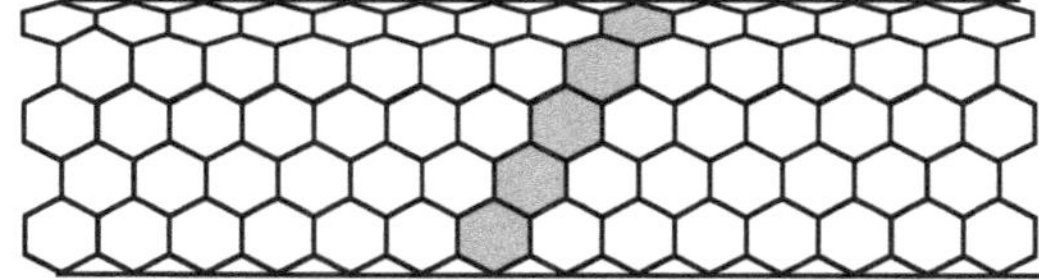

**(a) Zigzag nanotube**

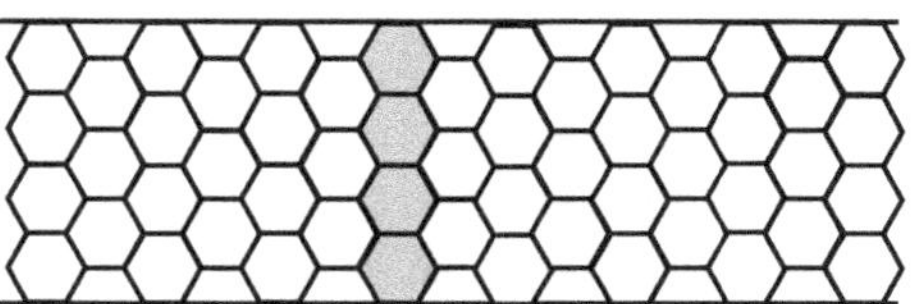

**(b) Armchair nanotube**

**Fig. 4.9**

- An infinite nanotube that is of the zigzag (or armchair) type consists entirely of closed zigzag (or armchair) paths, connected to each other.

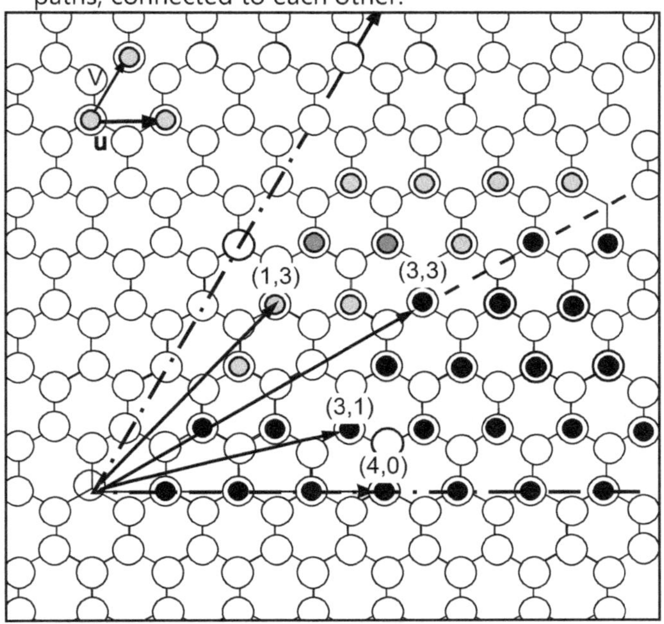


**Fig. 4.10 : The basis vectors *u* and *v* of the relevant sub-lattice, the (n,m) pairs that define non-isomorphic carbon nanotube structures (red dots), and the pairs that define the enantiomers of the chiral ones (blue dots).**

## Nanotube Types

- Moreover, the structure of the nanotube is not changed if the strip is rotated by 60 degrees clockwise around A1 before applying the hypothetical reconstruction above. Such a rotation changes the corresponding pair (n, m) to the pair (−2m, n+m).

- It follows that many possible positions of A2 relative to A1 — that is, many pairs (n, m) — correspond to the same arrangement of atoms on the nanotube. That is the case, for example, of the six pairs (1, 2), (−2, 3), (−3, 1), (−1,−2), (2,−3), and (3,−1). In particular, the pairs (k,0) and (0,k) describe the same nanotube geometry.

- These redundancies can be avoided by considering only pairs (n, m) such that n > 0 and m ≥ 0; that is, where the direction of the vector **w** lies between those of **u** (inclusive) and **v** (exclusive). It can be verified that every nanotube has exactly one pair (n,m) that satisfies those conditions, which is called the tube's **type**.

Conversely, for every type there is a hypothetical nanotube. In fact, two nanotubes have the same type if and only if one can be conceptually rotated and translated so as to match the other exactly.

- Instead of the type (n,m), the structure of a carbon nanotube can be specified by giving the length of the vector **w** (that is, the circumference of the nanotube), and the angle α between the directions of **u** and **w**, which may range from 0 (inclusive) to 60 degrees clockwise (exclusive). If the diagram is drawn with **u** horizontal, the latter is the tilt of the strip away from the vertical. Here are some unrolled nanotube diagrams:

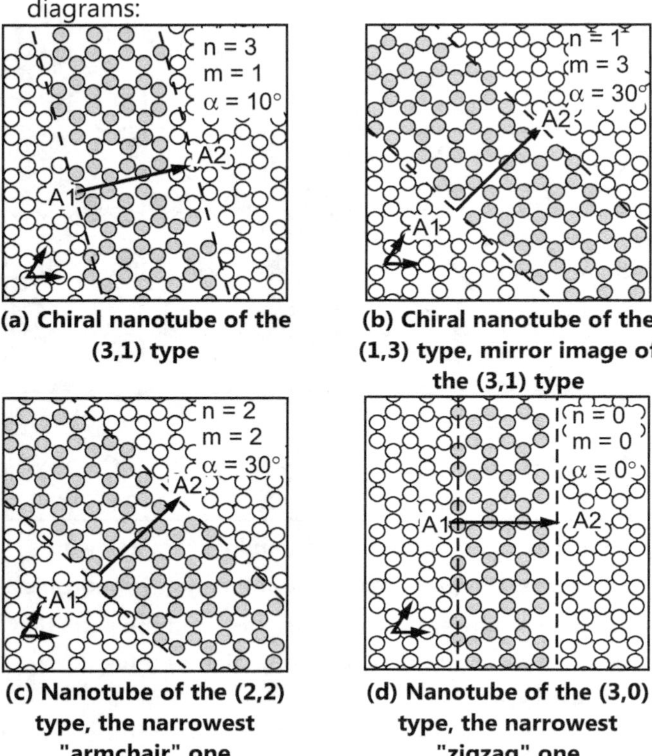

(a) Chiral nanotube of the (3,1) type

(b) Chiral nanotube of the (1,3) type, mirror image of the (3,1) type

(c) Nanotube of the (2,2) type, the narrowest "armchair" one

(d) Nanotube of the (3,0) type, the narrowest "zigzag" one

**Fig. 4.11 : Unrolled nanotube diagrams**

### Chiralty and Mirror Symmetry

- A nanotube is chiral if it has type (n,m), with m > 0 and m ≠ n; then its enantiomer (mirror image) has type (m,n), which is different from (n,m). This operation corresponds to mirroring the unrolled strip about the line L through A1 that makes an angle of 30 degrees clockwise from the direction of the **u** vector (that is, with the direction of the vector **u+v**). The only types of nanotubes that are achiral are the (k,0) "zigzag" tubes and the (k,k) "armchair" tubes.

- If two enantiomers are to be considered the same structure, then one may consider only types (n,m) with 0 ≤ m ≤ n and n > 0. Then the angle α between **u** and **w**, which may range from 0 to 30 degrees (inclusive both), is called the "chiral angle" of the nanotube.



- Based on the density functional theory combined with the nonequilibrium Green's function, the influence of the wrinkle on the electronic structures and transport properties of quasi-one-dimensional carbon nanomaterials have been investigated, in which the wrinkled armchair graphene nanoribbons (wAGNRs) and the composite of AGNRs and single walled carbon nanotubes (SWCNTs) were considered with different connection of ripples. The wrinkle adjusts the electronic structures and transport properties of AGNRs. With the change of the strain, the wAGNRs for three width families reveal different electrical behavior. The band gap of AGNR(6) increases in the presence of the wrinkle, which is opposite to that of AGNR(5) and AGNR(7). The transport of AGNRs with the widths 6 or 7 has been modified by the wrinkle, especially by the number of isolated ripples, but it is insensitive to the strain. The nanojunctions constructed by AGNRs and SWCNTs can form the quantum wells, and some specific states are confined in wAGNRs. Although these nanojunctions exhibit the metallic, they have poor conductance due to the wrinkle. The filling of $C_{20}$ into SWCNT has less influence on the electronic structure and transport of the junctions. The width and connection type of ripples have greatly influenced on the electronic structures and transport properties of quasi-one-dimensional nanomaterials.

- Graphene possesses unique physical and chemical properties, which has attracted significant interest. The wrinkle is a ubiquitous phenomenon in two dimensional (2D) membranes. Peierls and Landau *et al.* theoretically predicted that the 2D lattices can not exist at any finite temperatures, but it does not forbid nearly perfect 2D membranes from being embedded in 3D spaces. The existence of intrinsic wrinkle in a suspended graphene sheet was first discovered by Meyer *et al.* in 2007.

- Graphene sheets are easily wrinkled due to their relatively small bending rigidity and the compressive effect from substrate. The wrinkled graphene has been obtained by mechanical exfoliation, chemical vapor deposition, thermal load and local tension. Graphene wrinkle is categorized as simple ripple, standing collapsed and folded wrinkle. The wrinkle has a remarkable impact on the electrical structures and quantum transport properties of graphene. In a word, the presence of wrinkle leads to novel functions and properties of graphene, which is expected the potential applications in the transport of biological systems, separation science, and the development of the fluidic electronics.

- The electronic structure and transport properties of graphene can be strongly altered by ripple or protuberance. Much of previous investigations about wrinkle mainly focus on quasi-2D membranes, however, the wrinkled graphene nanoribbons remain less explored. In the presence of the wrinkle, a conductance gap is formed in armchair nanoribbons, which corresponds to a metal-semiconductor transition. With the fluctuation of wrinkle intensified, the overall averaged conductance decreases for both the zigzag and armchair nanoribbons. The transport through the ribbons depends sensitively on the bandstructure which, in turn, strongly depends on the geometry of the deformed ribbons. The effective potential is determined by the local curvature, altering significantly the transport properties of the ribbon. In this work, the influence of wrinkle on the electronic structures and transport properties of armchair graphene nanoribbons (AGRs) has been explored.

- The quasi-one-dimensional (quasi-1D) carbon nanomaterials are firstly considered for the wrinkled armchair graphene nanoribbons (wAGNRs) with one or two simple isolated ripples, in which two isolated ripples are connected in series. The electronic structure and transport properties of wAGNRs are further investigated by the change of the width and strain. Secondly, the influence of the wrinkle on the quasi-1D nanomaterials are investigated to the composite of AGNRs and single walled carbon nanotubes (SWCNTs). This system can be also formed by two wAGNRs joined with two AGNRs, where two wrinkled ripples are regarded as the connection in parallel.

- The fullerene could be encapsulated into a single wall carbon nanotube (SWCNT) to form the peapod system, which could result in unusual properties. Finally, this work will investigate the electronic structures and transport properties of the second system filled with fullerene $C_{20}$.

## 4.7.1 Calculation Methods

- Based on the density functional theory (DFT) combined with the non-equilibrium Green's function (NEGF), the electronic structure and transport properties of quasi-1D carbon nanomaterials have been implemented in the Atomistix ToolKit code package. The vacuum separation larger than 1.5 nanometer (nm) is used to eliminate the interaction between the nearest devices. A generalized gradient approximation (GGA) is used to describe the interaction between the valence electrons and the ion cores with the Perdew-Burke-Ernzerhof (PBE) as the exchange-correlation functional. The plane wave energy cutoff of 150 Rydberg is employed with the double zeta polarization basis set. The Brillouin zone k-point sampling is formed by the Monkhorst-Pack algorithm, and a 10×1×60 k-point grid is chosen for fully self-consistent calculations due to the protrusion in the x direction. All the structures are fully relaxed until the maximum force on each atom is less than 0.03 eV/Å during the optimization.The main feature of the DFT-NEGF computation is to simulate the transport of the whole two-probe system. The left and right lead voltages are set to $V_L = +V_b/2$ and $V_R = -V_b/2$ (where $V_b$ is the bias voltage), respectively.

- The creation of point defects in matter can profoundly affect the physical and chemical properties of materials. If appropriately controlled, these modifications can be exploited in applications promising advanced and novel functionalities. Redox-based memristive devices – one of the most attractive emerging memory technologies – provide one of the most striking examples for the potential exploitation of defects. Applying an external electric field to an initially insulating oxide layer is known to induce a non-volatile, voltage-history dependent switching between a low resistance state and a high resistance state, also named memristive device. This switching occurs through the creation and annihilation of the so-called conductive filaments, which are generated at the nanoscale by assembly of donor-type point defects such as oxygen vacancies.

- To date, the exact relationship between concentration and nanoscale distribution of defects within the filament on the one hand and the electronic transport properties of the devices on the other hand is still elusive. Due to limitations in sensitivity or spatial resolution of most characterization methods, the electronic structure of conductive filaments has not yet been characterized in detail. However, this knowledge is crucially needed as input for the development of electronic transport models with high predictive power.

- With this in mind, we characterized the electronic structure of epitaxial $SrTiO_{3-x}$ -based memristive devices and compare it to that of single crystalline $SrTiO_{3-x}$. To gain access to the defect states in the band gap, we employed soft X-ray resonant photoelectron spectroscopy (RESPES), which allows element-specific measurement of valence and defect levels of $SrTiO_{3-x}$ through resonant emission when the excitation energy corresponds to an absorption edge. The measurements were carried out in a photoelectron emission microscope (PEEM) setup at the Nanospectroscopy beamline of Elettra synchrotron laboratory, enabling us to spatially resolve the filament in the memristive device and presenting the first report of spatially resolved RESPES, a powerful tool to map the electronic structure of small features.

- In a first step, we investigated the spectroscopic fingerprint of oxygen vacancies in $SrTiO_{3-x}$ single crystals formed upon irradiation with an intense soft X-ray beam in ultra-high vacuum. We probed the occupied part of the electronic states near the Fermi energy using laterally-averaged RESPES. To this end, the valence band and the possibly occupied states within the band gap were mapped as a function of photon energy near the Ti L absorption edge (Fig. 4.12 a-b).

- Interestingly, we find occupied states of Ti 3d character between the Fermi energy and the valence band for the reduced case (light blue/green spots at photon energies ~458.9 eV and ~464.1 eV in Fig. 4.12 a)), which are absent in the oxidized case (Fig. 4.12 b)), suggesting that they are caused by the existence of oxygen vacancies. These in-gap states have relative maxima at ~0.31 eV and ~1.11 eV below the conduction band (Fig. 4.12 c-d)), and the photon energies at which they are observed indicate that these states are of $Ti^{3+}e_g$ character. These energy levels can be interpreted as the defect states for singly and doubly charged oxygen vacancies. Combined with the relative position of the Fermi level, these states are the key features of the electronic structure induced by a high density of oxygen vacancies in $SrTiO_{3-x}$.

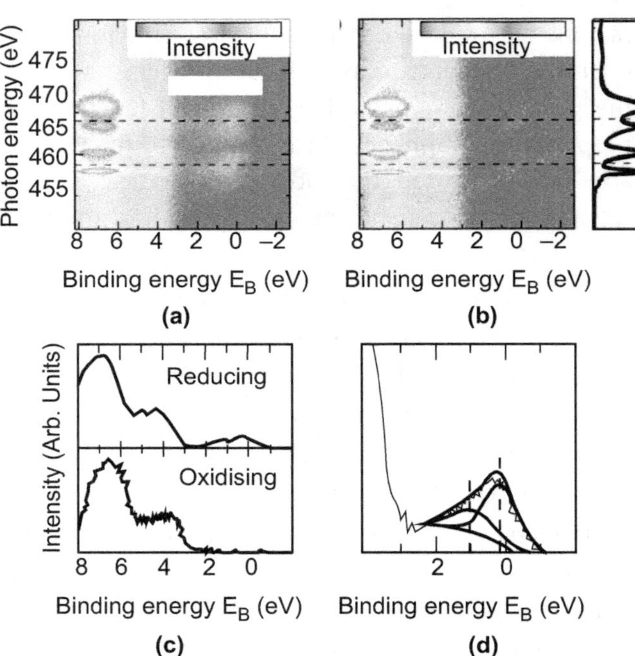


**Fig. 4.12 : RESPES analysis of $SrTiO_{3-x}$ single crystals. (a) RESPES map measured under reducing conditions. (b) RESPES map measured under oxidizing conditions and Ti L edge for reference. In both cases, second order light was subtracted from the RESPES map. (c) Valence band spectra at a photon energy of 463.3 eV. (d) Zoom-in for the valence band spectrum in reducing conditions**

- In order to gain access to the electronic structure of the conductive filament, we recorded PEEM images of the $SrTiO_{3-x}$ device near the Ti 3p core level and identified a well pronounced filament within the device area (Fig. 4.12 (a)). The conductive filament corresponds to reduced $SrTiO_{3-x}$ with up to 30 % $Ti^{3+}$, while the surrounding device area is composed almost entirely of $Ti^{4+}$ (Fig. 4.12 (a) and (b)). We mapped the valence band region in real space at a photon energy of 463.3 eV, an energy at which we could detect the in-gap states induced by the oxygen vacancies in the single crystalline reference. Again, we observe a clear contrast in the region of the conductive filament for the energies corresponding to the in-gap states (Fig. 4.12 (c)). Extracting spectra from the filament and the surrounding region reveals that only the former exhibits significant weight of the in-gap states in the spectra (Fig. 4.12 d)), whereas no in-gap states can be detected for the surrounding area. The in-gap states lie at the same energy positions compared to the valence band maximum as found for the reduced single crystalline reference (Fig. 4.12 (d)), indicating they are caused by the presence of oxygen vacancies.

## 4.8 APPLICATIONS TRANSPORT DEVICES

- Nano electronics holds some answers for how we might increase the capabilities of electronics devices while we reduce their weight and power consumption.

- Some of the nano electronics areas under development, which you can explore in more detail by following the links provided in the next section, include the following topics.

- Improving display screens on electronics devices. This involves reducing power consumption while decreasing the weight and thickness of the screens.

- Increasing the density of memory chips. Researchers are developing a type of memory chip with a projected density of one terabyte of memory per square inch or greater.

- Reducing the size of transistors used in integrated circuits. One researcher believes it may be possible to "put the power of all of today's present computers in the palm of your hand".

- Cadmium selenide nanocrystals deposited on plastic sheets have been shown to form Flexible Electronic Circuits. Researchers are aiming for a combination of flexibility, a simple fabrication process and low power requirements.

- Integrating Silicon nanophotonics components into CMOS integrated circuits. This optical technique is intended to provide higher speed data transmission between integrated circuits than is possible with electrical signals.

- Researchers at UC Berkeley have demonstrated a low power method to use Nanomagnets as switches, like transistors, in electrical circuits. Their method might lead to electrical circuits with much lower power consumption than transistor based circuits.

- Researchers at Georgia Tech, the University of Tokyo and Microsoft Research have developed a method to print prototype circuit boards using standard inkjet printers. Silver nanoparticle ink was used to form the conductive lines needed in circuit boards.

- Researchers at Caltech have demonstrated a laser that uses a nanopatterned silicon surface that helps produce the light with much tighter frequency control than previously achieved. This may allow much higher data rates for information transmission over fiber optics.

- Building transistors from carbon nanotubes to enable minimum transistor dimensions of a few nanometers and developing techniques to manufacture integrated circuits built with nanotube transistors.

- Researchers at Stanford University have demonstrated a method to make functioning integrated circuits using carbon nanotubes. In order to make the circuit work they developed methods to remove metallic nanotubes, leaving only semiconducting nanotubes, as well as an algorithm to deal with misaligned nanotubes. The demonstration circuit they fabricated in the university labs contains 178 functioning transistors.

- Developing a lead free solder reliable enough for space missions and other high stress environments using copper nanoparticles.

- Using electrodes made from Nanowires that would enable flat panel displays to be flexible as well as thinner than current flat panel displays.

- Using semiconductor nanowires to build transistors and integrated circuits.

- Transistors built in single atom thick graphene film to enable very high speed transistors.

- Researchers have developed an interesting method of forming PN junctions, a key component of transistors, in graphene. They patterned the p and n regions in the substrate. When the graphene film was applied to the substrate electrons were either added or taken from the graphene, depending upon the doping of the substrate. The researchers believe that this method reduces the disruption of the graphene lattice that can occur with other methods.

- Combining gold nanoparticles with organic molecules to create a transistor known as a NOMFET (Nanoparticle Organic Memory Field-Effect Transistor).

- Using carbon nanotubes to direct electrons to illuminate pixels, resulting in a lightweight, millimeter thick "nanoemmissive" display panel.

- Making integrated circuits with features that can be measured in nanometers (nm), such as the process that allows the production of integrated circuits with 22 nm wide transistor gates.

- Using nanosized magnetic rings to make Magnetoresistive Random Access Memory (MRAM).
  - ➤ Researchers have developed lower power, higher density method using nanoscale magnets called magnetoelectric random access memory (MeRAM).

- Developing molecular-sized transistors which may allow us to shrink the width of transistor gates to approximately one nm which will significantly increase transistor density in integrated circuits.

- Using self-aligning nanostructures to manufacture nanoscale integrated circuits.
- Using nanowires to build transistors without p-n junctions.
- Using buckyballs to build dense, low power memory devices.
- Using Magnetic Quantum Dots in spintronic semiconductor devices. Spintronic devices are expected to be significantly higher density and lower power consumption because they measure the spin of electronics to determine a 1 or 0, rather than measuring groups of electronics as done in current semiconductor devices.
- Using nanowires made of an alloy of iron and nickel to create dense memory devices. By applying a current magnetized sections along the length of the wire. As the magnetized sections move along the wire, the data is read by a stationary sensor. This method is called race track memory.
- Using silver nanowires embedded in a polymer to make conductive layers that can flex, without damaging the conductor.
- IMEC and Nantero are developing a memory chip that uses carbon nanotubes. This memory is labeled NRAM for Nanotube-Based Nonvolatile Random Access Memory and is intended to be used in place of high density Flash memory chips.
- Researcher have developed an organic nanoglue that forms a nanometer thick film between a computer chip and a heat sink. They report that using this nanoglue significantly increases the thermal conductance between the computer chip and the heat sink, which could help keep computer chips and other components cool.
- Researchers at Georgia Tech, the University of Tokyo and Microsoft Research have developed a method to print prototype circuit boards using standard inkjet printers. Silver nanoparticle ink was used to form the conductive lines needed in circuit boards.

## 4.9 2D SEMICONDUCTORS

- A two-dimensional semiconductor (also known as 2D semiconductor) is a type of natural semiconductor with thicknesses on the atomic scale. The rising research attention towards 2D semiconductors started with a discovery by Geim and Novoselov et al. in 2004, when they reported a new semiconducting material graphene, a flat monolayer of carbon atoms arranged in a 2D honeycomb lattice.

- A 2D monolayer semiconductor is significant because it exhibits stronger piezoelectric coupling than traditionally employed bulk forms, which enables 2D materials applications in new electronic components used for sensing and actuating. In this emergent field of research in solid-state physics, the main focus is currently on designing nanoelectronic components by the use of graphene as electrical conductor, hexagonal boron nitride as electrical insulator, and a transition metal dichalcogenide as semiconductor.

- 2-dimensional (2D) materials have been at the forefront of materials research in recent years due to their exotic electrical and optical properties and interesting mechanical properties deriving from their atomically thin dimensions. The isolation and synthesis of a range of atomically thin two-dimensional (2D) materials opened a new exciting platform to layer-by-layer materials and hybrid device engineering that enables the exploration and tailoring of superior or hitherto unknown properties and that promises a range of new technologies.

- This special feature is a focus issue on the science and applications that result from the inherent confinement in the out-of-plane direction in 2D materials. Thus, these materials present themselves as a surface, and this aspect differentiates them from other types of materials.

- The properties coming from this confinement need to be further addressed and, while many applications, including electronic, optoelectronic and photonic applications, are still in their infancy, the unique applications should be explored. This special issue will cover the cutting-edge issues, unresolved problems, and new results on material properties and device applications, including electronic and optical applications, of 2D materials, with a focus on the following topics:

**Growth, Characterization and Properties:**

- Formation of 2D-heterostructures and their physical properties
- Growth/deposition of 2D materials and their surface / electrical/optical characterization
- Optical/mechanical/chemical/electronic properties of 2D materials
- Elastic and mechanical properties of 2D systems
- Chemical properties of 2D materials

## 4.10 2D ELECTRONIC DEVICES

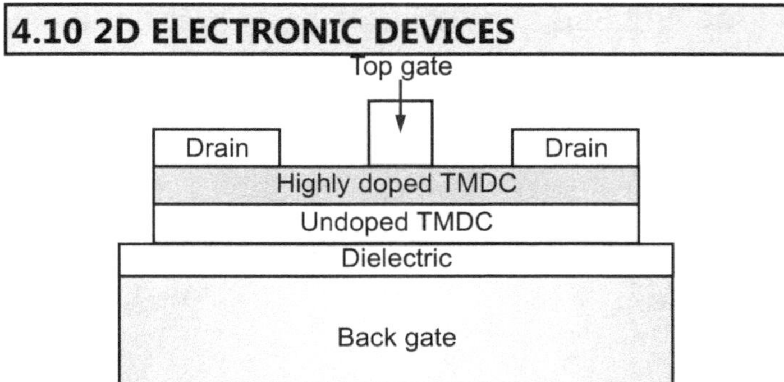


**Fig. 4.13**

- Field-effect transistors, tunneling transistors, flash memories and other devices based on 2D systems.
- Quantum transport and mesoscopic physics in 2D systems.
- Straintronics, Spintronics and Valleytronics with graphene and 2D systems.
- Thermal applications of 2D systems: heat-dissipation in nanocircuits.
- Mechanical and mechano-electrical devices based on 2D materials.
- Chemical, environmental, and biological sensor devices based on 2D materials.
- Magnetic tunnel junctions.

**Optoelectronic Devices:**
- Plasmonics, spintronics and valleytronics
- Lasers and light emitting diodes
- Photodetectors
- Photonic crystals and cavities
- Microwave, millimeter-wave, and terahertz devices
- Tunable optical metamaterials
- Quantum optics, nano-photonic devices
- Nonlinear optical properties and devices
- Flexible optoelectronic devices
- Opto-mechanical devices and system

Here, 2D materials include: graphene, h-BN, metal dichalcogenides, metal trichalcogenides, metal oxides, phosphorene, MXenes (Silicene, Germanene etc.), metal hallides, metal carbides, etc.

## 4.11 GRAPHENE

**Fig. 4.14 : Monolayer graphene**

- Graphene's two surfaces are single sheets of carbon atoms arranged in a hexagonal honeycomb lattice.

Having two surfaces and lacking bulk makes it the thinnest possible material but also 5 times stronger than steel due to pi and sigma orbital bonds. Graphene has high electron mobility and high thermal conductivity.

- Although graphene can be used in different applications, one issue regarding graphene is its lack of a band gap, which poses a problem in particular with digital electronics because it is unable to switch off field-effect transistors (FETs). Nanosheets of other group-IV elements (Si, Ge and Sn) present structural and electronic properties similar to graphene.

## 4.12 ATOMISTIC SIMULATION

- Atomistic simulations, the most widely used methods in the nanomechanics field, are important numerical methods for the investigation of magnetic, electronic, chemical, and mechanical properties of carbon nanostructures since these modeling approaches can accurately trace atomic position and precisely capture the microscale physical mechanism, such as buckling. There has been already much research of carbon nanostructures using atomistic simulation.

- Yakobson et al. studied the large deformation of CNTs using MD simulation. The change of each morphological pattern occurs accompanying an abrupt release of energy. Liew et al. investigated the thermal stability, the elastic and plastic properties of CNTs and twisting effect, as well as buckling on CNTs bundles using MD simulation which employs the second-generation of reactive empirical bond-order potential and Lennard–Jones potential to describe the atomic interaction. Liu et al. proposed an atomic-scale finite element method, which has the same framework as the traditional finite element method, for numerical simulation of CNTs. Feng and Liew employed MD simulation to analyze the stability and buckling of carbon nanorings by the definition of stability as no buckling occurring when mapping nanorings from CNTs. The temperature effect on the elastic properties has been examined by Hsieh et al.

- For the purpose of showing the promising application as nanoresonators and mass detectors, Li and Chou presented a molecular–structural–mechanics method to simulate vibration behaviors of CNT. Their results demonstrated that SWCNTs have ultrahigh fundamental frequency, a level of 10 GHz–1.5 THz for cantilevered or bridged boundary conditions. It was also observed that the fundamental frequencies of double-walled CNTs (DWCNTs) are 10% lower than those of SWCNTs of the same outer diameter and the noncoaxial vibration started at the third vibration mode. Chowdhury et al. adopted a molecular mechanics approach to investigate the vibration properties of two kinds of SWCNTs and found natural frequencies of zigzag CNTs are higher than those of the counterpart armchair cases. Hashemnia et al. determined the fundamental frequency of CNTs and graphene sheets by the means of molecular–structural–mechanics approach. It is observed that fundamental frequencies of CNTs are greater than those of graphene sheets. Reddy et al. employed atomistic simulations and continuum shell modeling to analyze free vibration of SWCNTs. They pointed out that a minimum potential configuration of CNT was important for the computation of the stiffness matrix.

- In spite of the precise trace of the atomic displacement of CNTs, these atomistic simulations are highly expensive in terms of computational resources. Therefore, their applications are extremely limited to a simple system with a small number of atoms, and do not meet the demand of engineering application. For examples, Liew et al. spent 36 h studying the buckling behavior of a (10, 10) SWCNT containing 2000 atoms with MD simulation in a single SGI origin 2000 CPU, whereas the investigation of a four-walled CNT containing 15,097 atoms required 4 months.

- Atomistic simulations have been widely used in the recent years to study the elementary mechanisms and interactions of dislocations. Atomistic simulations are a powerful tool for predicting dislocation-core structures on an atomic scale, but can suffer from serious artifacts, in particular depending on the determination of the interatomic potentials to which they are very sensitive. It is also often unrealistic to simulate thermally activated processes by standard molecular dynamics, which allow simulations on the $10^{-9}$ s time scale only. To avoid the latter issue, Rasmussen et al. (1997a, b) used a sophisticated method based on a configuration space-patch technique, called the nudged elastic band (NEB) method, which allows the determination of the minimum energy transition path of a dissociated screw dislocation from one glide plane to another. The clear advantage of this method is that there is no need to speculate on the cross-slip mechanism, which is an outcome of the simulation. Only the initial and final states of the dissociated screw dislocation need to be given. Selected configurations obtained with the same approach as Rasmussen et al. are shown in Fig. 4.15.

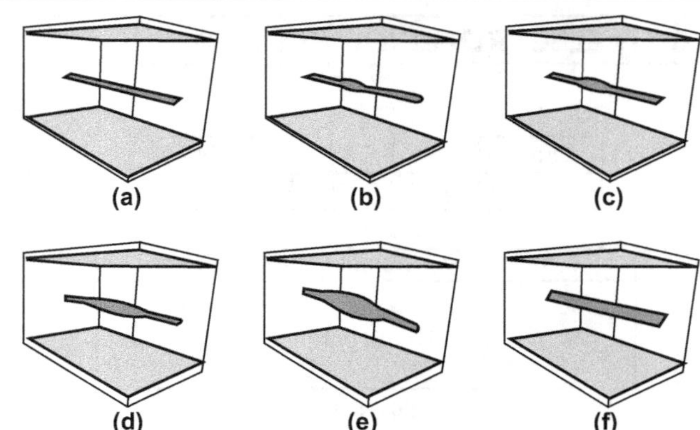


**Fig. 4.15 : Configurations of minimum energy path for cross slip according to atomistic calculations performed using the NEB method in copper. Only atoms with a non-FCC local environment.**

## SUMMARY

- Nanoelectronics refers to the use of nanotechnology in electronic components. The term covers a diverse set of devices and materials, with the common characteristic that they are so small that inter-atomic interactions and quantum mechanical properties need to be studied extensively.

- One type of RTDs is formed as a single quantum well structure surrounded by very thin layer barriers. This structure is called a double barrier structure.

- A Quantum dot is a nanostructure or tiny particle of a semiconducting substance with a diameter range of 2nm to 10nm only have discrete energy values inside the quantum well.

- Although using quantum dots as the basis for solar cells is not a new idea, attempts to make photovoltaic devices have not yet achieved sufficiently high efficiency in converting sunlight to power.

- Carbon nanotubes (CNTs) are tubes made of carbon with diameters typically measured in nanometers. Carbon nanotubes often refers to single-wall carbon nanotubes (SWCNTs) with diameters in the range of a nanometer.

## EXERCISE

1. Draw and explain Resonant Tunneling Diode.
2. Describe Coulomb dots.
3. Explain Quantum blockade.
4. Draw and explain Single electron transistors.
5. Explain Carbon nanotube electronics.
6. Describe Band structure and transport devices.
7. State Applications of transport devices.
8. What are the different 2D semiconductors.
9. Draw and explain 2D electronic devices.
10. Write a note on Graphene.
11. What is Atomistic simulation? Explain.

# PROPERTIES OF NANO DEVICES

## 5.1 INTRODUCTION

- After more than 20 years of basic nanoscience research and more than fifteen years of focused R&D under the NNI, applications of nanotechnology are delivering in both expected and unexpected ways on nanotechnology's promise to benefit society.

- Nanotechnology is helping to considerably improve, even revolutionize, many technology and industry sectors: information technology, homeland security, medicine, transportation, energy, food safety, and environmental science, among many others. Described below is a sampling of the rapidly growing list of benefits and applications of nanotechnology.

- Many benefits of nanotechnology depend on the fact that it is possible to tailor the structures of materials at extremely small scales to achieve specific properties, thus greatly extending the materials science toolkit. Using nanotechnology, materials can effectively be made stronger, lighter, more durable, more reactive, more sieve-like, or better electrical conductors, among many other traits. Many everyday commercial products are currently on the market and in daily use that rely on nanoscale materials and processes:

  - Nanoscale additives to or surface treatments of fabrics can provide lightweight ballistic energy deflection in personal body armor, or can help them resist wrinkling, staining, and bacterial growth.

  - Clear nanoscale films on eyeglasses, computer and camera displays, windows, and other surfaces can make them water- and residue-repellent, antireflective, self-cleaning, resistant to ultraviolet or infrared light, antifog, antimicrobial, scratch-resistant, or electrically conductive.

  - Nanoscale materials are beginning to enable washable, durable "smart fabrics" equipped with flexible nanoscale sensors and electronics with capabilities for health monitoring, solar energy capture, and energy harvesting through movement.

  - Lightweighting of cars, trucks, airplanes, boats, and space craft could lead to significant fuel savings. Nanoscale additives in polymer composite materials are being used in baseball bats, tennis rackets, bicycles, motorcycle helmets, automobile parts, luggage, and power tool housings, making them lightweight, stiff, durable, and resilient. Carbon nanotube sheets are now being produced for use in next-generation air vehicles. For example, the combination of light weight and conductivity makes them ideal for applications such as electromagnetic shielding and thermal management.

  - Nano-bioengineering of enzymes is aiming to enable conversion of cellulose from wood chips, corn stalks, unfertilized perennial grasses, etc., into ethanol for fuel. Cellulosic nanomaterials have demonstrated potential applications in a wide array of industrial sectors, including electronics, construction, packaging, food, energy, health care, automotive, and defense. Cellulosic nanomaterials are projected to be less expensive than many other nanomaterials and, among other characteristics, tout an impressive strength-to-weight ratio.

  - Nano-engineered materials in automotive products include high-power rechargeable battery systems; thermoelectric materials for temperature control; tires with lower rolling resistance; high-efficiency/low-cost sensors and electronics; thin-film smart solar panels; and fuel additives for cleaner exhaust and extended range.

  - Nanostructured ceramic coatings exhibit much greater toughness than conventional wear-resistant coatings for machine parts. Nanotechnology-enabled lubricants and engine oils also significantly reduce wear and tear, which can significantly extend the lifetimes of moving parts in everything from power tools to industrial machinery.

  - Nanoparticles are used increasingly in catalysis to boost chemical reactions. This reduces the quantity of catalytic materials necessary to produce desired results, saving money and reducing pollutants. Two big applications are in petroleum refining and in automotive catalytic converters.

  - Nano-engineered materials make superior household products such as degreasers and stain removers; environmental sensors, air purifiers, and filters; antibacterial cleansers; and specialized paints and sealing products, such a self-cleaning house paints that resist dirt and marks.

➤ Nanoscale materials are also being incorporated into a variety of personal care products to improve performance. Nanoscale titanium dioxide and zinc oxide have been used for years in sunscreen to provide protection from the sun while appearing invisible on the skin.

• Nanotechnology has greatly contributed to major advances in computing and electronics, leading to faster, smaller, and more portable systems that can manage and store larger and larger amounts of information. These continuously evolving applications include:

➤ Transistors, the basic switches that enable all modern computing, have gotten smaller and smaller through nanotechnology. At the turn of the century, a typical transistor was 130 to 250 nanometers in size. In 2014, Intel created a 14 nanometer transistor, then IBM created the first seven nanometer transistor in 2015, and then Lawrence Berkeley National Lab demonstrated a one nanometer transistor in 2016! Smaller, faster, and better transistors may mean that soon your computer's entire memory may be stored on a single tiny chip.

➤ Using Magnetic Random Access Memory (MRAM), computers will be able to "boot" almost instantly. MRAM is enabled by nanometer-scale magnetic tunnel junctions and can quickly and effectively save data during a system shutdown or enable resume-play features.

➤ Ultra-high definition displays and televisions are now being sold that use quantum dots to produce more vibrant colors while being more energy efficient.

➤ Flexible, bendable, foldable, rollable, and stretchable electronics are reaching into various sectors and are being integrated into a variety of products, including wearables, medical applications, aerospace applications, and the Internet of Things. Flexible electronics have been developed using, for example, semiconductor nanomembranes for applications in smartphone and e-reader displays. Other nanomaterials like graphene and cellulosic nanomaterials are being used for various types of flexible electronics to enable wearable and "tattoo" sensors, photovoltaic that can be sewn onto clothing, and electronic paper that can be rolled up. Making flat, flexible, lightweight, non-brittle, highly efficient electronics opens the door to countless smart products.

➤ Other computing and electronic products include Flash memory chips for smart phones and thumb drives; ultra-responsive hearing aids; antimicrobial/antibacterial coatings on keyboards and cell phone casings; conductive inks for printed electronics for RFID/smart cards/smart packaging; and flexible displays for e-book readers.

➤ Nanoparticle copper suspensions have been developed as a safer, cheaper, and more reliable alternative to lead-based solder and other hazardous materials commonly used to fuse electronics in the assembly process.

• Nanotechnology is already broadening the medical tools, knowledge, and therapies currently available to clinicians. Nanomedicine, the application of nanotechnology in medicine, draws on the natural scale of biological phenomena to produce precise solutions for disease prevention, diagnosis, and treatment. Below are some examples of recent advances in this area:

➤ Commercial applications have adapted gold nanoparticles as probes for the detection of targeted sequences of nucleic acids, and gold nanoparticles are also being clinically investigated as potential treatments for cancer and other diseases.

➤ Better imaging and diagnostic tools enabled by nanotechnology are paving the way for earlier diagnosis, more individualized treatment options, and better therapeutic success rates.

➤ Nanotechnology is being studied for both the diagnosis and treatment of atherosclerosis, or the buildup of plaque in arteries. In one technique, researchers created a nanoparticle that mimics the body's "good" cholesterol, known as HDL (high-density lipoprotein), which helps to shrink plaque.

➤ The design and engineering of advanced solid-state nanopore materials could allow for the development of novel gene sequencing technologies that enable single-molecule detection at low cost and high speed with minimal sample preparation and instrumentation.

- ➢ Nanotechnology researchers are working on a number of different therapeutics where a nanoparticle can encapsulate or otherwise help to deliver medication directly to cancer cells and minimize the risk of damage to healthy tissue. This has the potential to change the way doctors treat cancer and dramatically reduce the toxic effects of chemotherapy.

- ➢ Research in the use of nanotechnology for regenerative medicine spans several application areas, including bone and neural tissue engineering. For instance, novel materials can be engineered to mimic the crystal mineral structure of human bone or used as a restorative resin for dental applications. Researchers are looking for ways to grow complex tissues with the goal of one day growing human organs for transplant. Researchers are also studying ways to use graphene nanoribbons to help repair spinal cord injuries; preliminary research shows that neurons grow well on the conductive graphene surface.

- ➢ Nanomedicine researchers are looking at ways that nanotechnology can improve vaccines, including vaccine delivery without the use of needles. Researchers also are working to create a universal vaccine scaffold for the annual flu vaccine that would cover more strains and require fewer resources to develop each year.

- Nanotechnology is finding application in traditional energy sources and is greatly enhancing alternative energy approaches to help meet the world's increasing energy demands. Many scientists are looking into ways to develop clean, affordable, and renewable energy sources, along with means to reduce energy consumption and lessen toxicity burdens on the environment:

  - ➢ Nanotechnology is improving the efficiency of fuel production from raw petroleum materials through better catalysis. It is also enabling reduced fuel consumption in vehicles and power plants through higher-efficiency combustion and decreased friction.

  - ➢ Nanotechnology is also being applied to oil and gas extraction through, for example, the use of nanotechnology-enabled gas lift valves in offshore operations or the use of nanoparticles to detect microscopic down-well oil pipeline fractures.

  - ➢ Researchers are investigating carbon nanotube "scrubbers" and membranes to separate carbon dioxide from power plant exhaust.

- Researchers are developing wires containing carbon nanotubes that will have much lower resistance than the high-tension wires currently used in the electric grid, thus reducing transmission power loss.

  - ➢ Nanotechnology can be incorporated into solar panels to convert sunlight to electricity more efficiently, promising inexpensive solar power in the future. Nanostructured solar cells could be cheaper to manufacture and easier to install, since they can use print-like manufacturing processes and can be made in flexible rolls rather than discrete panels. Newer research suggests that future solar converters might even be "paintable."

  - ➢ Nanotechnology is already being used to develop many new kinds of batteries that are quicker-charging, more efficient, lighter weight, have a higher power density, and hold electrical charge longer.

  - ➢ An epoxy containing carbon nanotubes is being used to make windmill blades that are longer, stronger, and lighter-weight than other blades to increase the amount of electricity that windmills can generate.

  - ➢ In the area of energy harvesting, researchers are developing thin-film solar electric panels that can be fitted onto computer cases and flexible piezoelectric nanowires woven into clothing to generate usable energy on the go from light, friction, and/or body heat to power mobile electronic devices. Similarly, various nanoscience-based options are being pursued to convert waste heat in computers, automobiles, homes, power plants, etc., to usable electrical power.

  - ➢ Energy efficiency and energy saving products are increasing in number and types of application. In addition to those noted above, nanotechnology is enabling more efficient lighting systems; lighter and stronger vehicle chassis materials for the transportation sector; lower energy consumption in advanced electronics; and light-responsive smart coatings for glass.

- In addition to the ways that nanotechnology can help improve energy efficiency (see the section above), there are also many ways that it can help detect and clean up environmental contaminants:

  - ➢ Nanotechnology could help meet the need for affordable, clean drinking water through rapid, low-cost detection and treatment of impurities in water.

- ➤ Engineers have developed a thin film membrane with nanopores for energy-efficient desalination. This molybdenum disulphide ($MoS_2$) membrane filtered two to five times more water than current conventional filters.

- ➤ Nanoparticles are being developed to clean industrial water pollutants in ground water through chemical reactions that render the pollutants harmless. This process would cost less than methods that require pumping the water out of the ground for treatment.

- ➤ Researchers have developed a nanofabric "paper towel" woven from tiny wires of potassium manganese oxide that can absorb 20 times its weight in oil for cleanup applications. Researchers have also placed magnetic water-repellent nanoparticles in oil spills and used magnets to mechanically remove the oil from the water.

- ➤ Many airplane cabin and other types of air filters are nanotechnology-based filters that allow "mechanical filtration," in which the fiber material creates nanoscale pores that trap particles larger than the size of the pores. The filters also may contain charcoal layers that remove odors.

- ➤ Nanotechnology-enabled sensors and solutions are now able to detect and identify chemical or biological agents in the air and soil with much higher sensitivity than ever before. Researchers are investigating particles such as self-assembled monolayers on mesoporous supports (SAMMS™), dendrimers, and carbon nanotubes to determine how to apply their unique chemical and physical properties for various kinds of toxic site remediation. Another sensor has been developed by NASA as a smartphone extension that firefighters can use to monitor air quality around fires.

- Nanotechnology offers the promise of developing multifunctional materials that will contribute to building and maintaining lighter, safer, smarter, and more efficient vehicles, aircraft, spacecraft, and ships. In addition, nanotechnology offers various means to improve the transportation infrastructure:

  - ➤ As discussed above, nano-engineered materials in automotive products include polymer nanocomposites structural parts; high-power rechargeable battery systems; thermoelectric materials for temperature control; lower rolling-resistance tires; high-efficiency/low-cost sensors and electronics; thin-film smart solar panels; and fuel additives and improved catalytic converters for cleaner exhaust and extended range. Nano-engineering of aluminum, steel, asphalt, concrete and other cementitious materials, and their recycled forms offers great promise in terms of improving the performance, resiliency, and longevity of highway and transportation infrastructure components while reducing their life cycle cost. New systems may incorporate innovative capabilities into traditional infrastructure materials, such as self-repairing structures or the ability to generate or transmit energy.

- ➤ Nanoscale sensors and devices may provide cost-effective continuous monitoring of the structural integrity and performance of bridges, tunnels, rails, parking structures, and pavements over time. Nanoscale sensors, communications devices, and other innovations enabled by nanoelectronics can also support an enhanced transportation infrastructure that can communicate with vehicle-based systems to help drivers maintain lane position, avoid collisions, adjust travel routes to avoid congestion, and improve drivers' interfaces to onboard electronics.

- ➤ "Game changing" benefits from the use of nanotechnology-enabled lightweight, high-strength materials would apply to almost any transportation vehicle. For example, it has been estimated that reducing the weight of a commercial jet aircraft by 20 percent could reduce its fuel consumption by as much as 15 percent. A preliminary analysis performed for NASA has indicated that the development and use of advanced nanomaterials with twice the strength of conventional composites would reduce the gross weight of a launch vehicle by as much as 63 percent. Not only could this save a significant amount of energy needed to launch spacecraft into orbit, but it would also enable the development of single stage to orbit launch vehicles, further reducing launch costs, increasing mission reliability, and opening the door to alternative propulsion concepts.

## 5.2 VERTICAL TRANSISTORS

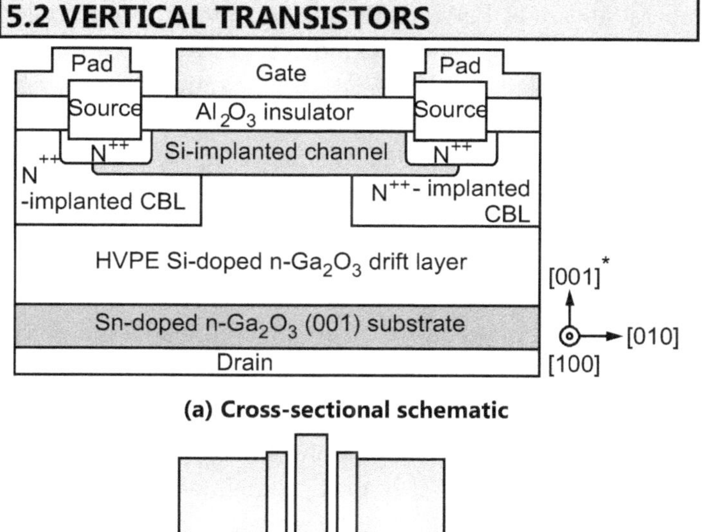


**(a) Cross-sectional schematic**



**(b) plan-view optical micrograph of the vertical**

**Fig. 5.1**

- Researchers at the National Institute of Information and Communications Technology (NICT) and Tokyo University of Agriculture and Technology (TUAT) demonstrate a vertical Ga$_2$O$_3$ metal-oxide-semiconductor field-effect transistor (MOSFET) that adopts an all-ion-implanted process for both n-type and p-type doping, paving the way for new generations of low-cost and highly manufacturable Ga$_2$O$_3$ power electronic devices.

- Power electronics is concerned with the regulation and conversion of electric power in such applications as motor drives, electric vehicles, data centers, and the grid. Power electronic devices, namely rectifiers (diodes) and switches (transistors), form the core components of power electronic circuits. Today, power devices made of silicon (Si) are the mainstream but they are approaching fundamental performance limitations, rendering the commercial power systems bulky and inefficient.

- A new generation of power devices based on the wide-bandgap semiconductor gallium oxide (Ga$_2$O$_3$) is expected to revolutionize the power electronics industry. Ga$_2$O$_3$ promises dramatic reductions in the size, weight, cost, and energy consumption of power systems by increasing both the power density and power conversion efficiency at the device level.

- The groundbreaking demonstration of the first single-crystal Ga$_2$O$_3$ transistor by NICT in 2011 galvanized intensive international research activities into the science and engineering of this new oxide semiconductor. For the past several years, the development of Ga$_2$O$_3$ transistors has focused on a lateral geometry. However, lateral devices are not amenable to the high currents and high voltages required for many applications owing to large device areas and reliability issues arising from self-heating and surface instabilities. In contrast, the vertical geometry allows for higher current drives without having to enlarge the chip size, simplified thermal management, and far superior field termination. The properties of a vertical transistor switch are engineered by introducing two types of impurities (dopants) into the semiconductor n-type doping, which provides mobile charge carriers (electrons) to carry electrical current when the switch is in the on-state; and p-type doping, which enables voltage blocking when the switch is in the off-state. A group at NICT led by Masataka Higashiwaki has pioneered the use of Si as an n-type dopant in Ga$_2$O$_3$ devices, but the community has long struggled to identify a suitable p-type dopant. Earlier this year, the same group published on the feasibility of nitrogen (N) as a p-type dopant. Their latest accomplishment involves integrating Si and N doping to engineer a Ga$_2$O$_3$ transistor for the first time, through a high energy dopant introduction process known as ion implantation.

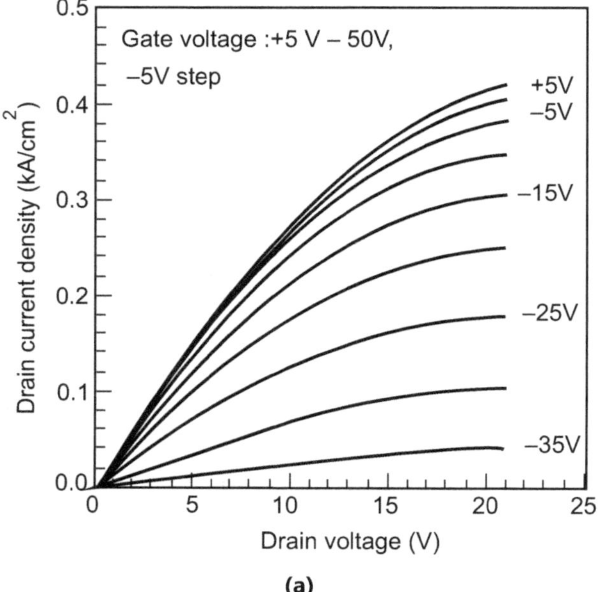


**(a)**

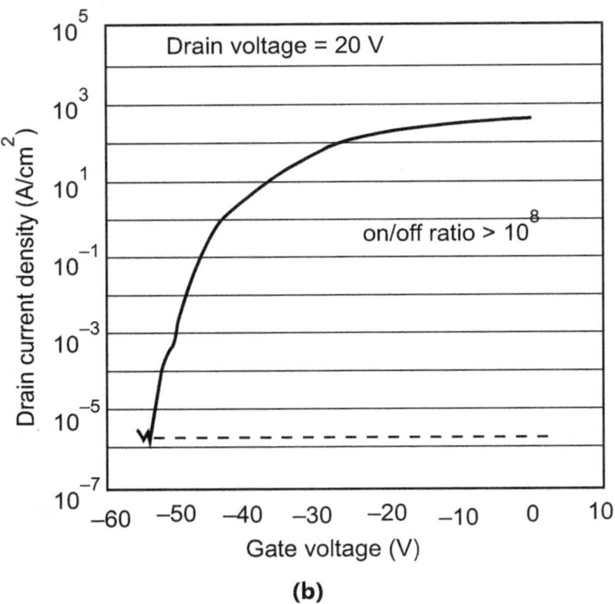


(b)

**Fig. 5.2 (a) DC output and (b) transfer characteristics of the vertical Ga$_2$O$_3$ MOSFET. Credit: National Institute of Information and Communications Technology (NICT)**

- "Our success is a breakthrough development that promises a transformational impact on Ga$_2$O$_3$ power device technology," said Higashiwaki, Director of the Green ICT Device Advanced Development Center at NICT. "Ion implantation is a versatile fabrication technique widely adopted in the mass production of commercial semiconductor devices such as Si and silicon carbide (SiC) MOSFETs. The demonstration of an all-ion-implanted vertical Ga$_2$O$_3$ transistor greatly enhances the prospects for Ga$_2$O$_3$-based power electronics."

- This study, published December 3 in the *IEEE Electron Device Letters* as an early access online paper and scheduled to appear in the January 2019 issue of the journal, builds on an earlier one in which a different acceptor dopant was used. "We initially investigated magnesium for p-type doping, but this dopant failed to deliver its expected performance since it diffuses significantly at high process temperatures," said Man Hoi Wong, a researcher of the Green ICT Device Advanced Development Center and the lead author of the paper. "Nitrogen, on the other hand, is much more thermally stable, thereby creating unique opportunities for designing and engineering a variety of high-voltage Ga$_2$O$_3$ devices."

- The Ga$_2$O$_3$ base material used for fabricating the vertical MOSFET was produced by a crystal growth technique called halide vapor phase epitaxy (HVPE). Pioneered by Profs. Yoshinao Kumagai and Hisashi

Murakami at TUAT, HVPE is capable of growing single-crystal Ga$_2$O$_3$ films at high speeds and with low impurity levels. Three ion implantation steps were performed to form the n-type contacts, n-type channel, and p-type current blocking layers (CBLs) in the MOSFET. The device showed decent electrical properties including an on-current density of 0.42 kA/cm$^2$, a specific on-resistance of 31.5 m$\Omega\cdot$cm$^2$, and a high drain current on/off ratio larger than eight orders of magnitude. Further improvements in its performance can be readily achieved with improved gate dielectric quality and optimized doping schemes.

- According to Higashiwaki and Wong, "Vertical power devices are the strongest contenders to combine currents over 100 A with voltages over 1 kV the requirements for many medium- and high-power industrial and automotive electric power systems." The technological impact of Ga$_2$O$_3$ will be substantially bolstered by the availability of melt-grown native substrates one of the key enablers of the silicon industry that dominates the global semiconductor market with an annual revenue of several hundred billion U.S. dollars. "The commercialization of vertical SiC and gallium nitride (GaN) power devices has, to a certain extent, been hindered by the high cost of substrates. For Ga$_2$O$_3$, the high quality and large size of native substrates offer this rapidly emerging technology a unique and significant cost advantage over the incumbent wide-bandgap SiC and GaN technologies".

  ➤ A lateral transistor uses the p-type layer for both the emitter and collector fabrications.

  ➤ A vertical transistor uses the p-type layer for emitter and the p-type substrate as collector. This transistor is sometimes known as substrate transistor.

  ➤ In both transistors, the base is made of the n-type epitaxial layer.

  ➤ Such transistors with a uniform and thick base are slow and the current gain $\beta$ of such transistors is small.

  ➤ When a PNP transistor with a large current gain is required, then the concept of the composite transistor (super-beta transistor) can be implemented.

- **Vertical Pnp:** The substrate of this structure serves as its collector, so that it can be used in applications where cathode is at common ground. It is used inoutput stages of op-amp.

- **Lateral PNP:** Tub is isolated from the substrate. Emitter current is lost by parasitic collection at the isolation wall and at the substrate. This results incurrent gain reduction.
- The lateral distances are usually longer than the vertical ones , so the transit times are also longer.
- The forward-biased emitter injects holes into the base under the emitter. Most of these holes are stored there, but part of them will be collected by the substrate, giving rise to a substrate current that reduces the gain.
- In comparison with vertical npn transistors, the lateral pnp transistors have a lower current gain (<50), low maximum transit frequencies (<100Mhz), and lower collector currents.

## 5.3 FIN FET

- A fin field-effect transistor (FinFET) is a multigate device, a MOSFET (metal-oxide-semiconductor field-effect transistor) built on a substrate where the gate is placed on two, three, or four sides of the channel or wrapped around the channel, forming a double gate structure. These devices have been given the generic name "finfets" because the source/drain region forms fins on the silicon surface. The FinFET devices have significantly faster switching times and higher current density than planar CMOS (complementary metal-oxide-semiconductor) technology.
- FinFET is a type of non-planar transistor, or "3D" transistor. It is the basis for modern nanoelectronic semiconductor device fabrication. Microchips utilizing FinFET gates first became commercialized in the first half of the 2010s, and became the dominant gate design at 14 nm, 10 nm and 7 nm process nodes.

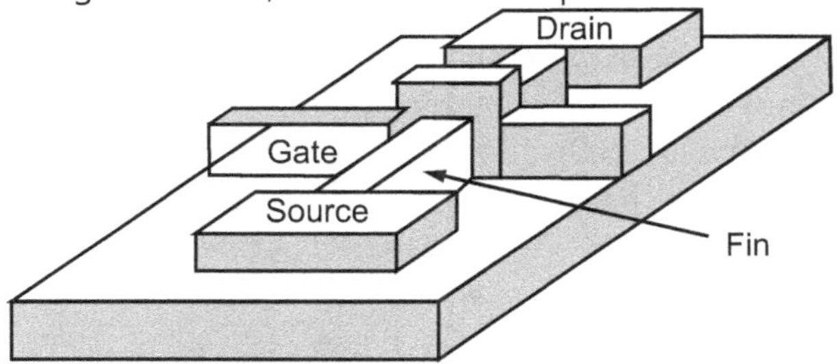


**Fig. 5.3 : A double-gate FinFET device**

- At the technical level, FinFET, or the fin field-effect transistor, is a particular kind of metal-oxide semiconductor transistor (MOSFET) that has a double or triple-gate structure that enables much faster operation and greater current density than traditional designs. That leads to lower voltage requirements, too, making the FinFET design far more energy efficient.
- Although the first FinFET transistor design was developed in the 1990s under the name of the Depleted Lean-channel Transistor, or DELTA transistor,

it wasn't until the early-2000s that the term FinFET was coined. It's an acronym of sorts, but the "fin" portion of the name was suggested because both the source and drain regions of the MOSFET form fins on the silicon surface it's built upon.

- The first commercial usage of FinFET technology was with the 25nm nanometer transistor created by TSMC in 2002. It was known as the "Omega FinFET" design, with further iterations on this technological idea coming in the years that followed, including Intel's Tri-Gate variant, which was introduced in 2011 with its 22nm Ivy Bridge microarchitecture.
- AMD also claimed to be working on similar technology in the early 2000s, although nothing really materialized from it; when AMD divested from its holdings in GlobalFoundries in 2009, the product and fabrications arms of the business were permanently severed.
- Starting in 2014, all major chip manufacturers GlobalFoundries included began using FiNFET technology based on 16nm and 14nm technology, eventually shrinking the node size to 7nm with the latest iterations.
- In 2019, additional technological advancements have allowed for even greater reductions in the length of FinFET gates, leading to 7nm. Within the next couple of years, even 5nm process technology for more powerful and efficient CPUs, graphics cards, and System on Chip (SoCs) will be achievable. However, these node sizes are approximate in most cases and not always directly comparable with TSMC and Samsung's latest 7nm technology, which is said to be roughly comparable to Intel's 10nm process.

### 5.3.1 Surround Gate FET

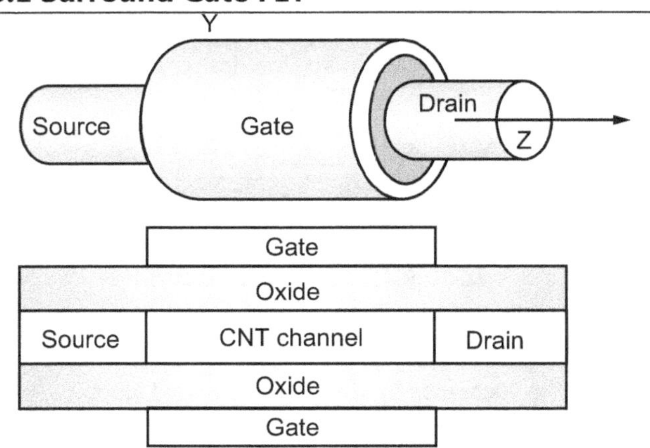


**Fig. 5.4 : Surround gate FET**

- Harnessing the potential of single crystal inorganic nanowires for practical advanced nanoscale applications requires not only reproducible synthesis of highly regular one-dimensional (1D) nanowire arrays directly on device platforms but also elegant device integration which retains structural integrity of the nanowires while significantly reducing or eliminating complex critical processing steps. Here we demonstrate a unique, direct, and bottom-up integration of a semiconductor 1D nanowire, using zinc oxide (ZnO) as an example, to obtain a vertical surround-gate field-effect transistor (VSG-FET).

- The vertical device structure and bottom-up integration reduce process complexity, compared to conventional top-down approaches. More significantly, scaling of the vertical channel length is lithographically independent and decoupled from the device packing density. A bottom electrical contact to the nanowire is uniquely provided by a heavily doped underlying lattice-match substrate. Based on the nanowire-integrated platform, both n- and p-channel VSG-FETs are fabricated. The vertical device architecture has the potential for use in tera-level ultrahigh-density nanoscale memory and logic devices.

## 5.4 METAL SOURCE/DRAIN JUNCTIONS

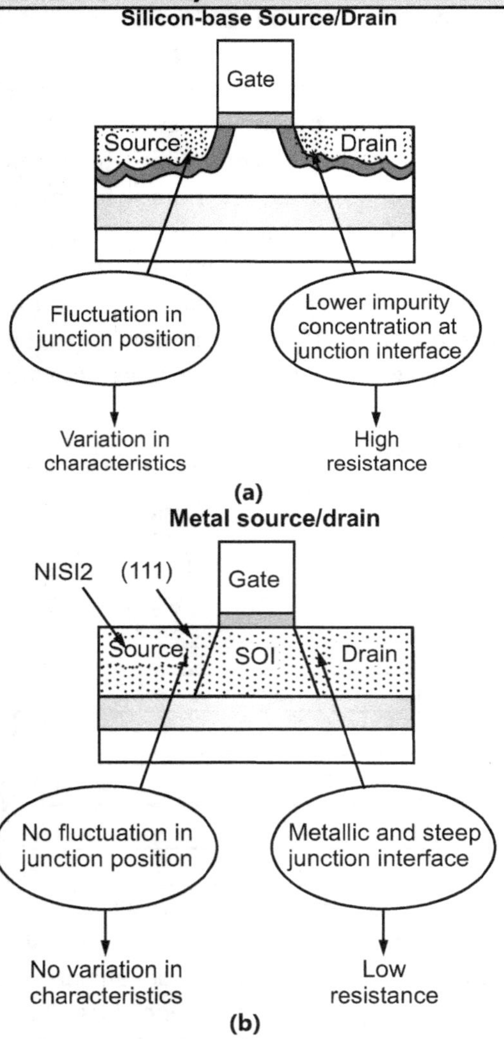

**Fig 5.5 : Comparison between the silicon semiconductor source–drain junction and the metal source–drain junction.**

- Above Fig. 5.5 compares a conventional heavily-doped silicon semiconductor source–drain junction and the metal source–drain junction. The silicon semiconductor source–drain junction is formed by ion implantation of phosphorus, arsenic, boron, and other impurities and high-temperature heat treatment. This results in variation in junction characteristics owing to fluctuation in implantation positions and thermal diffusion, and an increase in resistance due to a decrease in the concentrations of the dopants (phosphorus, arsenic, boron, and other impurities) at the junction interface. On the other hand, the metal source–drain junction is fabricated at a relatively low temperature by a solid-state reaction of a metal and silicon. This eliminates the variation in junction positions and creates steep junction interfaces, making it possible to significantly reduce the variation in junction characteristics.

- A $NiSi_2$ crystal tends to form a stable structure surrounded by (111) planes. We investigated the growth of a $NiSi_2$ crystal formed in a very thin silicon layer to examine the possibility of controlling the position of the metal source–drain junction in a very small transistor. An 8-nm-thick SOI layer was prepared and a dummy gate-stack structure was formed. A nickel film was formed and then heat treated at 500 °C to form a $NiSi_2$ crystal. As the size of this crystal is nearly the same as that of a silicon crystal, the $NiSi_2$ grows keeping an epitaxial structure. Unreacted nickel was removed and $HfO_2$ films were deposited to make the initial reaction rims clear in cross-sectional observation by a transmission electron microscope (TEM).

- Fig. 5.6 shows cross-sectional TEM images of the change in positions of the epitaxial $NiSi_2$ source–drain junctions with different heat treatment times. After 1 minute of heat treatment, the $NiSi_2$ crystal had metastable (100) and stable (111) planes. After 100 minutes of heat treatment, the crystal had grown in <100> direction, reached the embedded $SiO_2$ interface (BOX in the figure), and then stopped growing. The position of the (111) plane of $NiSi_2$ remained unchanged. After 300 minutes of heat treatment, the $NiSi_2$ crystal had grown laterally while maintaining the stable (111) plane. This growth corresponds to the change in the position of the transistor junction. The crystal grew laterally by 8 nm after the 300-minute heat treatment. In this phenomenon, the growth rate depends on the diffusion of atoms in the crystal, and therefore the extent of growth depends on the square root of the time. The growth rate was estimated to be 0.04 nm/min. Since the rate is very slow, the position of the junction can be controlled at the nanometer level.

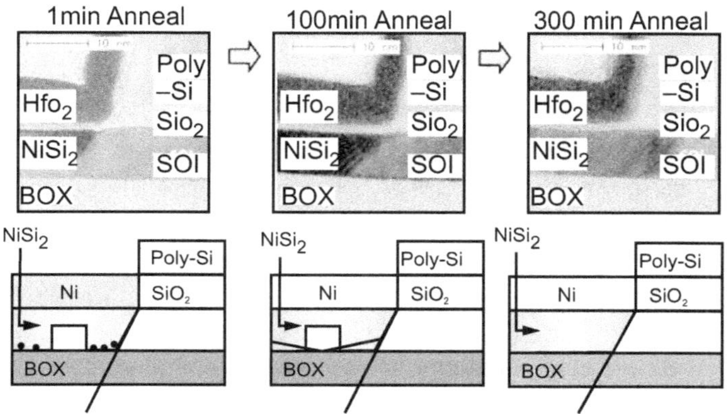



**Fig 5.6 : Cross-sectional TEM images showing the change in the positions of the epitaxial NiSi$_2$ source–drain junctions with different heat-treatment times. "BOX" in the figure denotes embedded SiO$_2$ and "poly-Si" denotes polycrystalline silicon.**

- We fabricated transistors with the same gate-stack structure and gate length by changing only the heat treatment time, and we then assessed the effect of the junction positions on transistor characteristics. Fig. 5.6 compares the characteristics of the transistors. The drain current increased by more than 20% when the junction positions were made closer. It was confirmed that the variation of the characteristics did not increase. This position control technology for the metal source–drain junction by utilizing the properties of NiSi$_2$ crystal growth is expected to be a new junction technology for the MOS transistors of 16-nm generation and beyond.

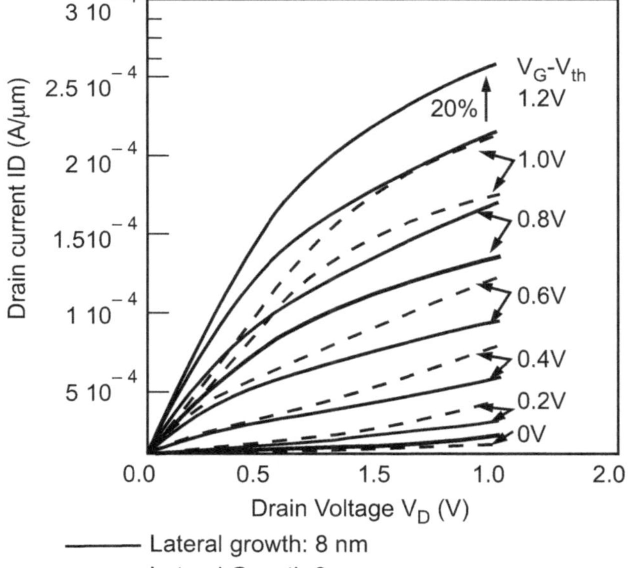



**Fig. 5.7 : Drain current–drain voltage (Id–Vd) characteristics of epitaxial NiSi$_2$ source–drain junction MOS transistors with a gate length of 50 nm: performance comparison between a transistor subjected to 1-minute heat treatment (0 nm lateral growth) and that subjected to 300-minute heat treatment  (8 nm lateral growth). The lateral growth of 8 nm meant that the effective gate length was reduced by 16 nm.**

## 5.5  PROPERTIES OF SCHOTTKY FUNCTIONS ON SILICON

There are several key Schottky properties which affect the functionality of Silicon or diode

- **Forward Voltage Drop:** In view of the low forward voltage drop across the diode, this is a parameter that is of particular concern. As can be seen from the Schottky diode I-V characteristic, the voltage across the diode varies according to the current being carried. Accordingly any specification given provides the forward voltage drop for a given current. Typically the turn-on voltage is assumed to be around 0.2 V.

- **Reverse Breakdown:** Schottky diodes do not have a high breakdown voltage. Figures relating to this include the maximum Peak Reverse Voltage, maximum Blocking DC Voltage and other similar parameter names. If these figures are exceeded then there is a possibility the diode will enter reverse breakdown. It should be noted that the RMS value for any voltage will be $1/\sqrt{2}$ times the constant value. The upper limit for reverse breakdown is not high when compared to normal PN junction diodes. Maximum figures, even for rectifier diodes only reach around 100 V. Schottky diode rectifiers seldom exceed this value because devices that would operate above this value even by moderate amounts would exhibit forward voltages equal to or greater than equivalent PN junction rectifiers.

- **Capacitance:** The capacitance parameter is one of great importance for small signal RF applications. Normally the junctions areas of Schottky diodes are small and therefore the capacitance is small. Typical values of a few picofarads are normal. As the capacitance is dependent upon any depletion areas, etc, the capacitance must be specified at a given voltage.

- **Reverse Recovery Time:** This parameter is important when a diode is used in a switching application. It is the time taken to switch the diode from its forward conducting or 'ON' state to the reverse 'OFF' state. The charge that flows within this time is referred to as the 'reverse recovery charge'. The time for this parameter for a Schottky diode is normally measured in nanoseconds, ns. Some exhibit times of 100 ps. In fact what little recovery time is required mainly arises from the capacitance rather than the majority carrier recombination. As a result there is very little reverse current overshoot when switching from the forward conducting state to the reverse blocking state.

- **Reverse Leakage Current:** The reverse leakage parameter can be an issue with Schottky diodes. It is found that increasing temperature significantly increases the reverse leakage current parameter. Typically for every 25°C increase in the diode junction temperature there is an increase in reverse current of an order of magnitude for the same level of reverse bias.

- **Working Temperature:** The maximum working temperature of the junction, Tj is normally limited to between 125 to 175°C. This is less than that which can be used with ordinary silicon diodes. Care should be taken to ensure heatsinking of power diodes does not allow this figure to be exceeded.

- **Limited Reverse Voltage:** As a result of its structure, Schottky diode rectifiers have a limited reverse voltage capability. The maximum figures are normally around 100 volts. If devices were manufactured with figures above this, it would be found that the forward voltages would rise and be equal to or greater than their equivalent silicon diodes for reasonable levels of current.

- **High Reverse Leakage Current:** Schottky diode rectifiers have a much higher reverse leakage current than standard PN junction silicon diodes. Although this may not be a problem in some designs it may have an impact on others.

- **Limited Junction Temperature:** The maximum junction temperature of a Schottky diode rectifier is normally limited to the range 125°C to 175°C but check the manufacturers ratings for the given component. This compares to temperatures of around 200°C for silicon diode rectifiers.

- **Adequate Heatsink:** Even though the Schottky diode offers a much lower forward voltage drop, in some power applications, significant levels of power can be dissipated. It is necessary to remember this and not assume that as the voltage drop is lower, that heatsinks will not be required. The maximum junction temperatures admissible are lower than for equivalent silicon diodes.

## 5.6 GERMANIUM AND COMPOUND SEMICONDUCTORS

- A semiconductor is a substance whose electrical conductivity lies between that of a conductor like copper or aluminum and an insulator like rubber or glass. It can conduct electricity under some conditions but insulate electricity under other conditions. Of the 92 elements existing on the earth, only several elements, including silicon (Si), germanium (Ge) and selenium (Se), can behave as semiconductors.

- In contrast to a semiconductor composed of a single element, one composed of two or more elements is called a compound semiconductor. Typical examples of compound semiconductors include gallium arsenide (GaAs), gallium nitride (GaN), indium phosphide (InP), zinc selenide (ZnSe), and silicon carbide (SiC).

**Single semiconductor vs compound semiconductor**

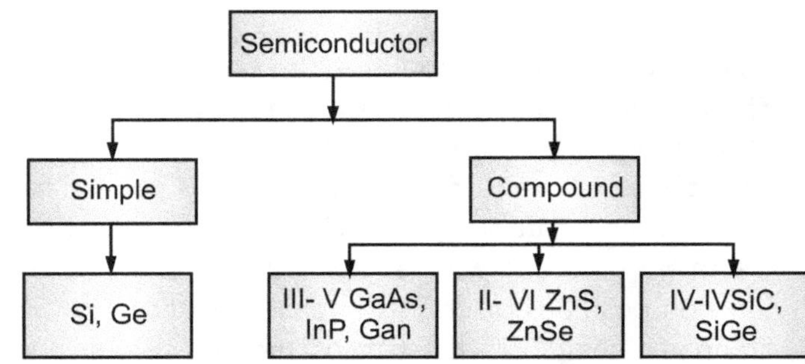


**Combinations on the periodic table of the elements**

| II | III | IV | V | VI |
|----|-----|-----|-----|-----|
|    | B | C | N |   |
|    | Al | Si | P |   |
| Zn | Ga | Ge | As | Se |
| Cd | In | Sn | Sb | Te |

**Fig. 5.8**

- Since electrons in compound semiconductors move much faster than electrons in silicon, the most popular single-element semiconductor, compound semiconductors are ideal for high-speed signal processing. They operate at a lower voltage, respond to light, and generate microwaves.

- Owing to their superior light receiving/emitting function, magnetic sensitivity and heat resistance, compound semiconductors are also used in various electronic systems/devices familiar to us. Some examples of such systems/devices include lasers and light-receiving elements for optical fiber communication, power amplifiers for mobile phones and other wireless communication devices, light sources for DVDs and Blu-ray discs, white LEDs for illumination, and solar batteries.

- Compound semiconductors are semiconductors that are made from two or more elements. Silicon is made from a single element, and therefore is not a compound semiconductor.

- Most compound semiconductors are from combinations of elements from Group III and Group V of the Periodic Table of the Elements (GaAs, GaP, InP and others). Other compound semiconductors are made from Groups II and VI (CdTe, ZnSe and others). It is also possible to use different elements from within the same group (IV), to make compound semiconductors such as SiC.

- In the past, compound semiconductors were not used in the widespread commercial applications and high production volumes typical for silicon. These crystals are more difficult to grow than silicon. The number of defects in the crystal is higher, and the cost of making the crystal is higher. Compound semiconductors also tend to be more fragile. All of these factors limited the growth of compound semiconductors for commercial use.

- In recent years, however, the cost of manufacturing compound semiconductors has come down. It is still much higher than silicon, but at the same time, the special properties of these crystals have become more important for certain applications. Because of their fundamental material properties, compound semiconductors can do things that simply aren't possible with silicon.

## 5.7 WORK FUNCTION PINNING

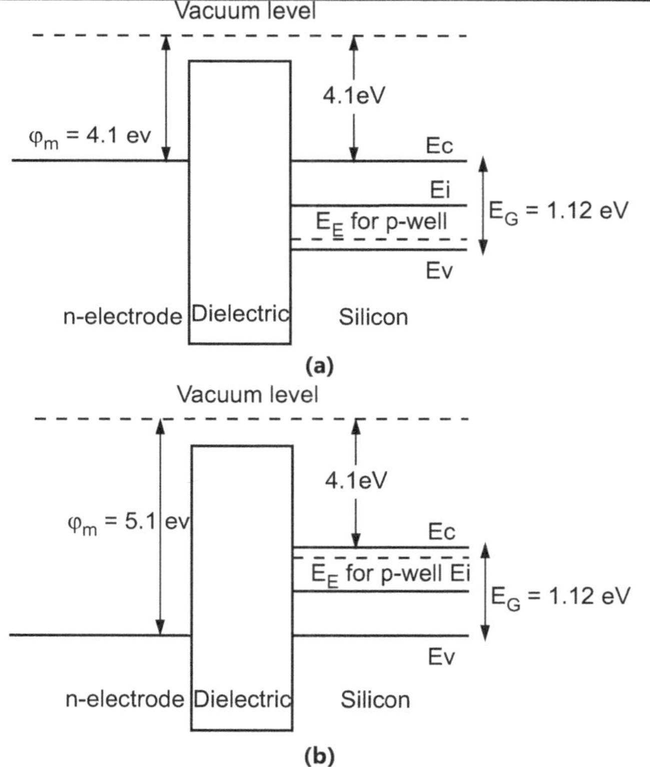


**Fig. 5.9 : Energy diagrams for NMOS and PMOS**
**φm is the work function of the gate electrode, EC the conduction band, EV the valence band, and EF the Fermi level energy**

- Deposition of monolayers of strong electron donors or acceptors on metal surfaces in many cases results in a metal-independent work function as a consequence of Fermi-level pinning. This raises the question whether in such a situation molecular dipoles, which are also frequently used to tune the interface energetics, still can have any impact. We use density functional theory to show that the spatial position of the dipoles is the determining factor and that only dipoles outside the immediate metal–molecule interface allow work-function changes beyond the pinning limit.

- Control of the work function of molybdenum disulfide ($MoS_2$) under ultra thin metal was investigated using in situ metal deposition and direct ultraviolet photoelectron spectroscopy measurement in an ultra-high vacuum system. When the metal thickness turned from two dimensional into bulk, the work function was also raised up at the nickel–$MoS_2$ interface, barely changed at the titanium–$MoS_2$ interface and lowered at the hafnium–$MoS_2$ interface. Meanwhile, the mechanisms of charge transfer and band alignment with metal deposition were also discussed. The Schottky barrier at metal–$MoS_2$ interfaces could be tailored by both types and thicknesses of deposited metal. The low work function metal was a good indicator for $MoS_2$ contact electrodes. It paved the way towards future high performance $MoS_2$ device applications.

- Recently, silicon carbide [SiC] has been proposed as the material of choice especially for power electronic and sensing devices operating under high temperature, fast switching, and high power conditions mainly due to its wide bandgap (3.26 eV), high critical electric field ($2.2 \times 10^6$ V/cm), superior thermal conductivity (4.9 W/Kcm), and high bulk electron mobility (900 $cm^2$/Vs) of the 4H polytype. For stable operations at high power densities and elevated temperatures, SiC diodes, including Schottky barrier diodes and junction barrier Schottky diodes, as well as SiC transistors, have been under extensive exploration with great improvements in wafer growth technology and device process.

- In order to realize stable SiC devices, metal contacts to SiC with suitable physical and electrical characteristics are required. For example, Ohmic contacts with low contact resistances and Schottky contacts with controlled barrier height ($\Phi_B$) between SiC and metal are among the most important factors for determining the performance of SiC devices. Furthermore, electrical characteristics of devices, such as voltage drop and switching speed of such devices, are dependent on the current transport behavior through the structure of the metal/4H-SiC interface. It is, therefore, of critical importance to reduce the barrier height of the metal/4H-SiC interface in order to improve the on-state voltage drop in 4H-SiC devices.

- To date, extensive studies have been carried out on the properties of barrier height of various metals on n- and p-types for SiC, and many attempts have been made to modify the contact barrier height on SiC. The effect of in homogeneities and Fermi-level pinning on Schottky contact properties has been known to be minimal, and the barrier height depends mostly on the metal work function without strong Fermi-level pinning for SiC . Recent work on the electrical contacts to SiC includes the implementation of nanostructures such as metal nanoparticles [NPs] to modify the barrier height at metal-SiC interfaces and to alter fundamental SiC device properties by controlling the size of the metal NPs. Previous results in the literature have been primarily focused on the effect of size reduction of NPs on the characteristics of diode structures with embedded NPs, which experimentally investigates the change in transport properties of metal/semiconductor interfaces in SiC depending on the size of NPs. However, so far, the focus has been mainly on the scaling effect of the NPs rather than on altering the electrical barrier of the NPs.

- In this work, we demonstrate that the work function change in the embedded metal NPs can effectively control the barrier height change of the SiC diode structures. Our results show that incorporating NPs with a larger work function difference to the capping metal layer results in an improved barrier lowering by further enhancing the local electric field. The experimental results have also been compared with both an analytic model based on Tung's theory and physics-based two-dimensional numerical simulations.

- The work function change in the embedded metal NPs can effectively lower the barrier height of the SiC diode structures. It has been experimentally shown that incorporating NPs (Ag) with a larger work function difference to the capping metal layer (Ni) results in an improved barrier lowering by further enhancing the local electric field. The barrier height of the fabricated SiC diode structures (NP-1; Ni with embedded Ag-NPs) has significantly reduced by 0.11 eV and 0.18 eV with respect to the samples with Au-NPs (NP-2) and the reference samples, respectively. The experimental results are in agreement with both analytic calculations based on Tung's model and physics-based two-dimensional numerical simulations, which confirm that

the increased electric field of the samples with NPs is mainly attributed to the reduction of barrier height as the effective barrier of the conduction band at the depletion region of the surface decreases.

## SUMMARY

- Nanotechnology is helping to considerably improve, even revolutionize, many technology and industry sectors: information technology, homeland security, medicine, transportation, energy, food safety, and environmental science, among many others.

- Nano-engineered materials in automotive products include high-power rechargeable battery systems; thermoelectric materials for temperature control; tires with lower rolling resistance; high-efficiency/low-cost sensors and electronics

- Nanotechnology is being studied for both the diagnosis and treatment of atherosclerosis, or the buildup of plaque in arteries

- Nanotechnology enabled sensors and solutions are now able to detect and identify chemical or biological agents in the air and soil with much higher sensitivity than ever before.

- A lateral transistor uses the p-type layer for both the emitter and collector fabrications.

- A vertical transistor uses the p-type layer for emitter and the p-type substrateas collector. This transistor is sometimes known as substrate transistor

- A fin field-effect transistor (FinFET) is a multigate device, a MOSFET (metal – oxide - semiconductor field effect transistor) built on a substrate where the gate is placed on two, three, or four sides of the channel or wrapped around the channel, forming a double gate structure.

- Forward voltage drop: In view of the low forward voltage drop across the diode, this is a parameter that is of particular concern. As can be seen from the Schottky diode I-V characteristic, the voltage across the diode varies according to the current being carried.

## EXERCISE

1. Draw and explain Vertical transistors.
2. Draw and explain the operation of Fin FET.
3. Draw and explain the operation of Surround g.
5. Explain the working of Metal source / drain junctions
6. What are the different Properties of schottky functions on Silicon ?
7. Explain Germanium and compound semiconductors.
8. Describe Work function pinning.

## UNIT VI

# CHARACTERIZATION TECHNIQUES FOR NANO MATERIALS

## 6.1 INTRODUCTION

- Nanoscale materials often present properties different from their bulk counterparts, as their high surface-to-volume ratio results in an exponential increase of the reactivity at the molecular level. Such properties include electronic, optical and chemical properties, while the mechanical characteristics of the nanoparticles (NPs) may also differ extensively.

- This enables them to be an object of intensive studies due to their academic interest and the prospective technological applications in various fields. Such nanostructures may be synthesized by a wide number of methods, which involve mechanical, chemical and other pathways. Nowadays, many more types of nanomaterials are synthesized than only a decade ago, and in higher amounts than before, requiring the development of more precise and credible protocols for their characterization.

- However, such characterization is sometimes incomplete. This is because of the inherent difficulties of nanoscale materials to be properly analysed, compared to the bulk materials (e.g. too small size and low quantity in some cases following laboratory-scale production). In addition, the multidisciplinary aspects of nanoscience and nanotechnology do not permit every research team to have easy access to a broad range of characterization facilities. In fact, quite often a wider characterization of NPs is necessary, requiring a comprehensive approach, by combining techniques in a complementary way.

- In this context, it is desirable to know the limitations and strengths of the different techniques, in order to know if in some cases the use of only one or two of them is enough to provide reliable information when studying a specific parameter (e.g. particle size). Nanoscience and nanotechnology are still undergoing constant growth, and the scientific community is rather aware that there may be certain differences between the way analytical characterization methods operate for nanomaterials, in comparison with their more 'traditional' modes of use for more 'conventional' (macroscopic) materials.

- Herein we describe extensively the use of different methods for the characterization of NPs. These techniques are sometimes exclusive for the study of a particular property, while in other cases they are combined. We discuss all these techniques in a comparative way, considering factors such as their availability, cost, selectivity, precision, non-destructive nature, simplicity and affinity to certain compositions or materials.

- The techniques are analysed in depth, despite their big number presented herein. There are microscopy-based techniques (e.g. TEM, HRTEM, and AFM – the full names of the techniques are provided later in the text, when presenting each one of them), which provide information on the size, morphology and crystal structure of the nanomaterials. Other techniques are specialized for certain groups of materials, such as the magnetic techniques.

- Examples of these techniques are SQUID, VSM, FMR, and XMCD. Many other techniques provide further information on the structure, elemental composition, optical properties and other common and more specific physical properties of the nanoparticle samples. Examples of these techniques include X-ray, spectroscopy and scattering techniques.

- This review is organized in different sections, which will present numerous distinct characterization techniques for NPs in relation to the properties studied (see Tables 6.1 and 6.2). The sections are categorized according to the different technique groups, as described above.

Table 6.1 Summary of the experimental techniques that are used for nanoparticle characterization featured in this paper

**Table 6.1**

| Technique | Main Information Derived |
|---|---|
| XRD (group: X-ray based techniques) | Crystal structure, composition, crystalline grain size |
| XAS (EXAFS, XANES) | X-ray absorption coefficient (element-specific) – chemical state of species, interatomic distances, Debye–Waller factors, also for non-crystalline NPs |
| SAXS | Particle size, size distribution, growth kinetics |

| Technique | Main Information Derived |
|---|---|
| XPS | Electronic structure, elemental composition, oxidation states, ligand binding (surface-sensitive) |
| FTIR (group: further techniques for structure/composition/main properties) | Surface composition, ligand binding |
| NMR (all types) | Ligand density and arrangement, electronic core structure, atomic composition, influence of ligands on NP shape, NP size |
| BET | Surface area |
| TGA | Mass and composition of stabilizers |
| LEIS | Thickness and chemical composition of self-assembled monolayers of NPs |
| UV-Vis | Optical properties, size, concentration, agglomeration state, hints on NP shape |
| PL spectroscopy | Optical properties – relation to structure features such as defects, size, composition |
| DLS | Hydrodynamic size, detection of agglomerates |
| NTA | NP size and size distribution |
| DCS | NP size and size distribution |
| ICP-MS | Elemental composition, size, size distribution, NP concentration |
| SIMS, ToF-SIMS, MALDI | Chemical information (surface-sensitive) on functional group, molecular orientation and conformation, surface topography, MALDI for NP size |
| RMM-MEMS, $\zeta$-potential, pH, EPM, GPC, DSC, etc. | Please check the relevant parts of the manuscript |
| SQUID-nanoSQUID (group: magnetic nanomaterials) | Magnetization saturation, magnetization remanence, blocking temperature |
| VSM | Similar to SQUID through M–H plots and ZFC-FC curves |

| Technique | Main Information Derived |
|---|---|
| Mössbauer | Oxidation state, symmetry, surface spins, magnetic ordering of Fe atoms, magnetic anisotropy energy, thermal unblocking, distinguish between iron oxides |
| FMR | NP size, size distribution, shape, crystallographic imperfection, surface composition, M values, magnetic anisotropic constant, demagnetization field |
| XMCD | Site symmetry and magnetic moments of transition metal ions in ferro- and ferri-magnetic materials, element specific |
| Magnetic susceptibility, magnetophoretic mobility | Please check the relevant parts of the manuscript |
| Superparamagnetic relaxometry | Core properties, hydrodynamic size distribution, detect and localize superparamagnetic NPs |
| TEM (group: microscopy techniques) | NP size, size monodispersity, shape, aggregation state, detect and localize/quantify NPs in matrices, study growth kinetics |
| HRTEM | All information by conventional TEM but also on the crystal structure of single particles. Distinguish monocrystalline, polycrystalline and amorphous NPs. Study defects |
| Liquid TEM | Depict NP growth in real time, study growth mechanism, single particle motion, superlattice formation |
| Cryo-TEM | Study complex growth mechanisms, aggregation pathways, good for molecular biology and colloid chemistry to avoid the presence of artefacts or destroyed samples |
| Electron diffraction | Crystal structure, lattice parameters, study order–disorder transformation, long-range order parameters |

| Technique | Main Information Derived |
|---|---|
| STEM | Combined with HAADF, EDX for morphology study, crystal structure, elemental composition. Study the atomic structure of hetero-interfaces |
| Aberration-corrected (STEM, TEM) | Atomic structure of NP clusters, especially bimetallic ones, as a function of composition, alloy homogeneity, phase segregation |
| EELS (EELS-STEM) | Type and quantity of atoms present, chemical state of atoms, collective interactions of atoms with neighbors, bulk plasmon resonance |
| Electron tomography | Realistic 3D particle visualization, snapshots, video, quantitative information down to the atomic scale |
| SEM-HRSEM, T-SEM-EDX | Morphology, dispersion of NPs in cells and other matrices/supports, precision in lateral dimensions of NPs, quick examination–elemental composition |
| EBSD | Structure, crystal orientation and phase of materials in SEM. Examine microstructures, reveal texture, defects, grain morphology, deformation |
| AFM | NP size and shape in 3D mode, evaluate degree of covering of a surface with NP morphology, dispersion of NPs in cells and other matrices/supports, precision in lateral dimensions of NPs, quick examination–elemental composition |
| MFM | Standard AFM imaging together with the information of magnetic moments of single NPs. Study magnetic NPs in the interior of cells. Discriminate from non-magnetic NPs |

Table 6.2 Parameters needed to be determined and the corresponding characterization techniques.

### Table 6.2

| Entity Characterized | Characterization Techniques Suitable |
|---|---|
| Size (structural properties) | TEM, XRD, DLS, NTA, SAXS, HRTEM, SEM, AFM, EXAFS, FMR, DCS, ICP-MS, UV-Vis, MALDI, NMR, TRPS, EPLS, magnetic susceptibility |
| Shape | TEM, HRTEM, AFM, EPLS, FMR, 3D-tomography |
| Elemental-chemical composition | XRD, XPS, ICP-MS, ICP-OES, SEM-EDX, NMR, MFM, LEIS |
| Crystal structure | XRD, EXAFS, HRTEM, electron diffraction, STEM |
| Size distribution | DCS, DLS, SAXS, NTA, ICP-MS, FMR, superparamagnetic relaxometry, DTA, TRPS, SEM |
| Chemical state–oxidation state | XAS, EELS, XPS, Mössbauer |
| Growth kinetics | SAXS, NMR, TEM, cryo-TEM, liquid-TEM |
| Ligand binding / composition / density / arrangement / mass, surface composition | XPS, FTIR, NMR, SIMS, FMR, TGA, SANS |
| Surface area, specific surface area | BET, liquid NMR |
| Surface charge | Zeta potential, EPM |
| Concentration | ICP-MS, UV-Vis, RMM-MEMS, PTA, DCS, TRPS |
| Agglomeration state | Zeta potential, DLS, DCS, UV-Vis, SEM, Cryo-TEM, TEM |
| Density | DCS, RMM-MEMS |
| Single particle properties | Sp-ICP-MS, MFM, HRTEM, liquid TEM |
| 3D visualization | 3D-tomography, AFM, SEM |
| Dispersion of NP in matrices / supports | SEM, AFM, TEM |
| Structural defects | HRTEM, EBSD |
| Detection of NPs | TEM, SEM, STEM, EBSD, magnetic susceptibility |
| Optical properties | UV-Vis-NIR, PL, EELS-STEM |
| Magnetic properties | SQUID, VSM, Mossbauer, M |

## 6.2 FTIR (FOURIER TRANSFORM INFRARED SPECTROSCOPY)

- FTIR is one techniques that help in the determination of the structure, composition, size and other basic features of the NPs. Fourier Transform Infrared Spectroscopy (FTIR) is a technique based on the measurement of the absorption of electromagnetic radiation with wavelengths within the mid-infrared region (4000–400 $cm^{-1}$).

- If a molecule absorbs IR radiation, the dipole moment is somehow modified and the molecule becomes IR active. A recorded spectrum gives the position of bands related to the strength and nature of bonds, and specific functional groups, providing thus information concerning molecular structures and interactions. Feliu and co-workers studied how Pt nanostructures performed on ethanol oxidation, using a combined approach with *in situ* ATR-FTIR and differential electrochemical mass spectroscopy (DEMS).

- These techniques helped to probe adsorbates electrochemically and detect volatile reaction products. Their results were in agreement with previous findings, showing that the preferred decomposition products were related to surface structures, with $CO_{ads}$ formation on (100) domains and acetaldehyde/acetic acid formation on (111) domains. In another report, carbon-supported platinum NPs (3–8 nm size) were used for the CO oxidation reaction and this catalytic process was monitored using DRIFTS and quadrupole mass spectrometry (QMS).

- The FTIR measurements of adsorbed CO confirmed the variations of $CO_{ad}$ and $O_{ad}$ in different steps of the experiment, in accordance with the results from QMS, while modifications in the CO distributing over various types of Pt surface sites were also noticed. Overall, DRIFTS was regarded as an important tool for the probation of the surface structure of Pt NPs under in situ conditions. Shukla et al. published a paper devoted to the FTIR investigation of the surfactant bonding to FePt NPs that were synthesized in the presence of oleic acid and oleylamine. The former ligand was found to bond to FePt NPs in both monodentate and bidentate forms, while oleylamine bonded to FePt molecularly with the $NH_2$ group intact.

- Furthermore, Au/Ag bimetallic NPs stabilized with dodecanethiol and soluble in nonpolar solvents were produced through a two-phase synthetic route in water/toluene mixtures. The most important insight derived from XPS and FTIR measurements was that Ag atoms were enriched at the outer part of the alloy clusters in comparison with the Au atoms. In another work, the influence of the Ag NP content on the photocatalytic degradation of oxalic acid adsorbed on $TiO_2$ NPs was evaluated by ATR-FTIR.

- Various Ag NP amounts were tested, and it was demonstrated that the incorporation of only a small quantity (2%) boosted the photocatalytic performance of $TiO_2$ NPs substantially. AFM and XPS were used to characterize the topography and chemical structure / composition of the composite NP films. Tzitzios *et al.* synthesized Ni NPs with a hexagonal crystal structure in the size range of 13–25 nm *via* the reduction of nickel stearate in the presence of PEG, oleic acid and oleylamine.

- FTIR spectra showed the presence of the characteristic groups at the surface of the NPs, such as the –HC=CH– arrangement in OAc and OAm, while the binding modes of the ligands onto the NP surface were also examined. Copper zinc tin sulpho-selenide ($CZTS_xSe_{1-x}$) nanocrystals were prepared by Haram and colleagues with a hot-injection process. The precursors were dissolved in OAm and heated at $T > 200°C$ for the synthesis. FTIR measurements showed the adsorption of OAm onto the surface of the particles. Characteristic bands arising from the moieties existing at the molecule of OAm and indicating its successful coordination with the NPs were spotted.

- Superparamagnetic ferrite NPs ($MFe_2O_4$, M = Ni, Co, Zn, Mn) with high crystallinity and size below 10 nm were synthesized by Sabale *et al.* with a simple 'polyol' method. The observation of tetrahedral (v1) frequency at the FTIR spectrum verified the presence of the spinel ferrite structure. Bands assigned to the –OH and C–O groups denoted the presence of diethylene glycol, thus revealing its successful coating around the ferrite NPs, endowing them a high solubility in water.

- In another report, the presence of Fe–O–P bonds was shown by FTIR measurements for hydrophobic iron nanoparticles which were treated with alkyl phosphonic acid-based ligands in order to turn them into water-soluble iron–iron oxide core–shell NPs. FTIR spectroscopy was also employed to characterize multifunctional $Fe_3O_4$@C@Ag hybrid NPs, which were prepared with a facile route based on the direct adsorption and spontaneous reduction of Ag ions onto the surface shell of C-coated magnetic NPs.

- The presence of carboxyl and hydroxyl groups on the NP surface was shown by FTIR. Certain bands attributed to carboxyl vibration implied that carbonyl and other reductive groups were oxidized by Ag ions, which was an indirect sign for the presence of Ag in the products. These hybrid nanostructures displayed a remarkable photocatalytic activity for the photodegradation of neutral red dye under visible light irradiation.

- Duong *et al.* have shown that FTIR can be successfully used to assess the affinity of polymers bearing phosphate groups as surface ligands for $NaYF_4$:Yb/Er upconversion nanoparticles. Trioctylphosphine-stabilized CdS nanorods synthesized by Chen *et al.* were also characterized by FTIR, revealing C–P stretching peaks related to the aforementioned molecule. Table 6.3 presents the IR vibrational assignments of several characteristic functional groups which are involved in nanoparticle synthesis.

Table 6.3 Selected infrared vibrational assignments for some of the most common groups present on the surface of NPs.

**Table 6.3**

| Vibrational Modes | Frequency ($cm^{-1}$) |
|---|---|
| Methyl C–H asym/sym stretch | 2970–2950/2880–2860 |
| Methyl C–H asym/sym bend | 1470–1430/1380–1370 |
| C=C alkenyl stretch | 1680–1620 |
| Aromatic C–H stretch | 3130–3070 |
| O–H hydroxyl group, H-bonded OH stretch | 3570–3200 (broad) |
| C–O stretch, primary alcohol | ~1050 |
| N–H aliphatic primary amine, NH stretch | 3400–3380, 3345–3325 |
| N–H primary amine, NH bend | 1650–1590 |
| C–N, primary amine, CN stretch | 1090–1020 |
| Carboxylate | 1610–1550/1420–1300 |
| Organic phosphates (P=O stretch) | 1350–1250 |
| Aliphatic phosphates (P–O–C stretch) | 1050–990 |
| Sulfonates | 1365–1340/1200–1100 |
| Organic siloxane or silicone (Si–O–Si) | 1095–1075/1055–1020 |
| Organic siloxane or silicone (Si–O–C) | 1100–1080 |
| Thiols (S–H stretch) | 2600–2550 |
| Thiol or thioether, $CH_2$–S–(C–S stretch) | 710–685 |
| Aliphatic chloro-compounds, C–Cl stretch | 800–700 |
| Ammonium ion | 3300–3030/1430–1390 |

# 6.3 XRD (X-RAY-BASED TECHNIQUES)

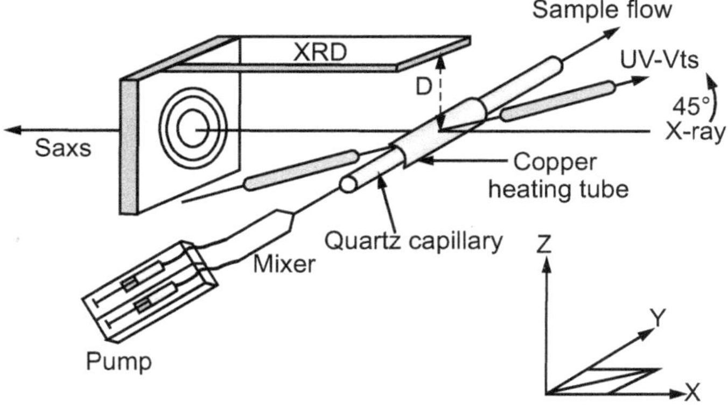


**Fig. 6.1**

- Fig. 6.1 Schematic presentation of the in situ setup employed for real-time SAXS / WAXS / UV–Vis measurements during the formation of Au NPs. The setup measures SAXS, WAXS and the UV–Vis spectra simultaneously in the same sample volume.

- X-ray diffraction (XRD) is one of the most extensively used techniques for the characterization of NPs. Typically, XRD provides information regarding the crystalline structure, nature of the phase, lattice parameters and crystalline grain size. The latter parameter is estimated by using the Scherrer equation using the broadening of the most intense peak of an XRD measurement for a specific sample.

- An advantage of the XRD techniques, commonly performed in samples of powder form, usually after drying their corresponding colloidal solutions, is that it results in statistically representative, volume-averaged values. The composition of the particles can be determined by comparing the position and intensity of the peaks with the reference patterns available from the International Centre for Diffraction Data (ICDD, previously known as Joint Committee on Powder Diffraction Standards, JCPDS) database. However, it is not suitable for amorphous materials and the XRD peaks are too broad for particles with a size below 3 nm.

- Upadhyay *et al.* determined the average crystallite size of magnetite NPs using X-ray line broadening, and it was found to be in the range of 9–53 nm. The broadening of XRD peaks was mainly caused by particle/crystallite size and lattice strains other than instrumental broadening. The XRD-derived size is usually bigger than the so-called magnetic size, due to the fact that smaller domains are present in a particle where all moments are aligned in the same direction, even if the particle is single domain.

- On the contrary, the TEM-deduced size was higher than that calculated using XRD, for samples with very large particles; in fact, when the particle size is bigger than 50 nm, there are more than one crystal boundary on their surface. XRD cannot distinguish between the two boundaries; therefore the actual (TEM) size of certain samples can be in reality bigger than the 50–55 nm calculated by the Scherrer formula.

- Dai and co-workers prepared ultra-small Au NPs which were very likely to be more developed along the $\langle 111 \rangle$ direction (rather than the $\langle 220 \rangle$ one) as the peak corresponding to the former direction was much more intense in their XRD measurement. Similarly, Li and colleagues noticed that after preparing copper telluride nanostructures with different shapes (i.e. cubes, plates, and rods), the relative intensities between the different XRD peaks varied in relation to the particle shape.

- Moroz reviewed the X-ray diffraction structure diagnostics of nanomaterials and stated that a remarkable advantage of EXAFS over REDD (radial electron density distribution) is its selectivity, whereas REDD is better in providing accurate values of the interatomic distances; in that case, EXAFS provides interatomic distances corrected for the phase shift.

- Ideally, REDD should be combined with EXAFS, FTIR and microscopy techniques to acquire knowledge on the relation between the structure and physicochemical properties of nanomaterials. In another work, Gomes *et al.* combined XRD and EXAFS to determine the cation distribution and other structural parameters, comparing the NP-based sample spectrum with the standard bulk material spectrum of the Cu ferrite.

- Differences were found among the cation redistribution at the nanoparticle samples with regard to the ideal copper ferrite. $CeO_2$ NPs were characterized with EXAFS by Zhu and co-workers. The authors emphasized on the suitability of the technique under discussion for their materials due to its element selectivity and nondependence on the long-range order of materials.

- From the acquired Debye–Waller factors and the Ce–O bond lengths, it was deduced that the surface or interface of the NPs coated with sodium bis(2-ethylhexyl) sulfosuccinate (AOT) surfactants was quite ordered; however the bond lengths were elongated. Swatsitang and colleagues analysed with EXAFS the impact of cation distribution on the magnetic properties of $Co_{1-x}Ni_xFe_2O_4$ NPs prepared by a hydrothermal method.

- The results implied that Co and Ni ions could occupy both the tetrahedral and octahedral sites with the preference to occupy the octahedral site more than the tetrahedral site, which is different from the bulk sample where all cations occupy only the octahedral site in an inverse spinel ferrite model structure.

- Furthermore, Zhang *et al.* synthesized $Co@SiO_2$ core-shell NPs with the sol–gel approach. *In situ* XRD was used along with EXAFS to monitor the oxidation process of the Co cores after thermal treatment at 800 °C either in air or under an inert atmosphere. Interestingly, it was noticed that Co was oxidized in three steps no matter if air or $N_2$ gas was employed during the annealing.

- Ni–P was another material in nanoscale form studied by EXAFS. In particular, EXAFS proved to be very robust for the screening of the initial crystallization behaviour of such amorphous NPs by probing the atomic-level structural change. Its combination with XRD, HRTEM and VSM helped to investigate in detail the structural changes of Ni–P NPs in both short-range and long-range order during heating at high temperatures. More specifically, XRD illustrated the crystalline phases and phase changes.

- HRTEM provided information on size, size distribution and shape. EXAFS provided insights regarding the changes of a local atomic structure and the chemical valence, especially for XRD-amorphous samples. VSM enabled the study of magnetic properties corresponding to different crystallization stages.

- Metal chalcogenides have also been analysed by EXAFS, as in the case of CdS NPs prepared by Rockenberger *et al.* They found EXAFS to be suitable for their samples since it does not rely on any long-range order, in contrast to XRD. EXAFS can also be used for liquid samples or even in cluster beams in the gaseous phase, permitting the identification of intercluster interactions by comparison with solid state measurements.

- It revealed that the stabilization of CdS NPs with 1.3–12 nm diameter affected the mean Cd–S distance. Unlike XRD, EXAFS is only sensitive to the local geometrical arrangement of neighboring atoms that surround the absorbing atom. O'Brien and colleagues investigated the local environment of phosphorus in the capping agent on the surface of CdSe quantum dots. The binding mode of the capping agents onto the surface was analysed, depending on the use of two distinct synthetic routes followed for its preparation (ligands: trioctylphosphine oxide and / or trioctylphosphine selenide).

- Furthermore, Lloyd and co-workers used various techniques, including EXAFS, to monitor the reduction of Se(IV) to Se(II) by a microbial whole-cell catalyst (*Veillonella atypica*). The reduction was found to proceed *via* an insoluble red amorphous Se(0) phase and the formation of metal selenide was shown by EXAFS analysis from both *ex situ* and *in situ* ways.

- Noble metal nanostructures, either monometallic or bimetallic, have also been studied by EXAFS. For example, the structural features of silver NPs embedded in silicate glass were investigated combining HRTEM and EXAFS techniques. Bugaev's group employed HRTEM, XRD, optical absorption and EXAFS to identify correlations between the plasmonic properties and the atomic structure of Ag NPs and their aggregates. The processing of the Ag K-edge EXAFS spectra resulted in the acquisition of values for the parameters of the atomic structure in Ag–Ag and Ag–O bonds averaged over the ionic and neutral states of Ag.

- The determination of the atomic structure of metallic Ag NPs, such as the type of point symmetry in the interior (core) region of small NPs, the nearest-neighbor Ag–Ag distances and the structure of the near-surface region, is a challenging problem for this system. The same group analysed by EXAFS the changes in the atomic structure of Ag NPs in soda-lime glass after annealing at 550 °C for 8 h. Brunsch and co-workers found that EXAFS showed a higher accuracy than HRTEM in the determination of lattice parameters for Ag NPs embedded in silicate glasses.

- The EXAFS results were averaged parameters of atom–atom correlations summed up for all detected particles, different from the data derived by HRTEM for single particles. Combining both techniques would be beneficial for precise structural characterization. The same researchers achieved the determination of the

- thermal expansion coefficient of similar glass-embedded silver NPs, in a wide range of temperatures. EXAFS allowed the precise demonstration of the changes of bond lengths and the stress state of their materials on the basis of a thermal expansion mismatch.

- The evaluation of the EXAFS data of small NPs typically provides a decreased coordination number, a dilatation or a contraction of the lattice structure and an increased Debye–Waller factor, together with an increased static disorder.

- In addition, EXAFS does not require operation under vacuum, in contrast to XPS. It was used by the Parkin group to investigate the phase change in silver speciation, during the photo-assisted growth of Ag from $AgNO_3$. The most plausible transition from metallic silver to $Ag_2O$ was illustrated.

- EXAFS has also been applied for the characterization of alloys such as $Pd_xPt_y$ NPs since common analytical techniques, such as XRD, have a limited capacity to distinguish between the various compositions of the aforementioned alloy, *i.e.* these metals are perfectly miscible in any relative proportion and they both possess an fcc structure with similar lattice constants. Furthermore, structural information can also be obtained from EXAFS measurements, even though sometimes with lower precision and difficulty in extraction in comparison with the use of XRD.

- Ingham has written a comprehensive review describing what X-ray scattering techniques such as EXAFS, *in situ* XRD and *small-angle X-ray scattering* (SAXS) can offer in nanoparticle characterization. Beale and Weckhuysen have reported how the ratio between coordination numbers varied as a function of shape, through the EXAFS data.

- Their study concerned a series of nanoscale structures with several shapes and fcc, hcp, or bcc structures, with a maximum isotropic diameter of 3 nm. In the case of Pt–Ru nanoclusters, size could also be obtained from EXAFS analysis, due to the fact that the coordination number of nearest neighbors in NPs is a non-linear function of the particle diameter if the latter parameter lies below the range of 3–5 nm.

- Sokolov and co-workers characterized their Pd NPs – synthesized by two separate routes – by X-ray reflectivity, EXAFS and electron microscopy. The EXAFS-deduced size was lower than the TEM one if a cuboctahedral fcc structural model of Pd NPs surrounded by thiol was assumed. Compared to bulk

Pd, lattice expansion was noticed in all types of NPs, by both HRTEM and EXAFS. Regarding cobalt NPs, a report by Cheng *et al.* showed that EXAFS was able to differentiate ε-Co from the fcc and hcp crystal structures.

- The combination of XANES and EXAFS for the characterization of CuO, $Cu_2O$/CuO and CuO/$TiO_2$ NPs was published by Sharma *et al.* These techniques probed the local electronic/atomic structure of these samples and the existence of the different oxide phases was evidenced. Iron oxide NPs were also studied by XANES: this technique provided information on the oxidation state and local structure of iron atoms, while SAXS analysis was useful for the size determination of the particles.

- The combination of time-resolved *in situ* SAXS and XANES measurements enabled the study of the formation of maghemite NPs in water, on a structural and chemical level. On the basis of the acquired data, a complex four-stage formation mechanism was proposed, using ferrous and ferric chlorides as well as triethanolamine.

- The acquired knowledge would facilitate the control of the formation of NPs in solution and tailor the properties of the final product. In another publication, Leveneur *et al.* investigated the nucleation and growth of Fe NPs in $SiO_2$ by TEM, XPS and XANES. It was demonstrated that ion implantation initially resulted in the formation of dilute cationic $Fe^{2+}$ species, while at higher dissolved iron concentrations, the formation of small metallic nuclei was noticed, which seed the nanocluster growth during prolonged implantation or annealing.

- XANES is a technique far more sensitive to coordination and bonding environment than XPS since it probes the unoccupied electronic states of atoms and therefore can provide information about the crystal field (octahedral, square pyramidal or tetrahedral) that the iron cations occupy. The complementary use of both XPS and XANES was considered to be handy for such types of nanostructures with complex compositions and various possible states for the valence of iron.

- Similar to the XRD method presented above, the SAXS technique allows elastic scattering processes into a given solid angle to be run; however the detector in SAXS covers only small scattering angles (normally lower than 1°). A scheme that illustrates an *in situ* setup which manages to record real-time SAXS/WAXS/UV–Vis measurements during the formation of Au NPs is displayed in Fig. 6.1. The pictured device conducts SAXS and WAXS and records the UV–Vis spectra at the same time in a given sample volume.

- WAXS (wide-angle X-ray scattering) is similar to SAXS, but the distance between the sample and the detector is smaller and therefore diffraction maxima at larger angles are observed. The authors investigated the nucleation and growth kinetics of gold NPs as a function of parameters such as concentration, temperature, ligand ratio and solvent type.

- Stuhn and co-workers characterized extensively polystyrene-grafted $SiO_2$ NPs, using techniques such as SAXS, SANS, DLS, TGA and TEM. Small angle neutron scattering (SANS) provided direct access to the static structure of the polymer layer. Both SAXS and SANS can be used to measure the particle size; in that report, SAXS gave a 25.6 nm value, while a 23.3 nm NP size was derived by SANS.

- Although SANS and SAXS are very similar in various aspects (*e.g.* SANS uses elastic neutron scattering), the advantages of SANS over SAXS include its sensitivity to light elements, the possibility of isotope labelling and the strong scattering by magnetic moments. The ligand shells on small ZnO NPs were characterized by a combined SANS/SAXS approach.

- Standard *in situ* methods such as UV-Vis and SAXS are sensitive only to the ZnO core; however SANS probes the organic stabilizer in dispersion thanks to the high sensitivity of neutrons to $H_2$. In the work under discussion, both techniques allowed the determination of the size distribution of the cores of the NPs and the distribution of the stabilizer molecules (acetate shell) simultaneously in the native solution.

## 6.4 AFM (ATOMIC FORCE MICROSCOPY)

- Atomic force microscopy is a microscopy technique capable of creating three-dimensional images of surfaces at high magnification. It was initially developed by Gerard Binning and Heinrich Rohrer at IBM in 1986. AFM is based on measuring the interacting forces between a fine probe and the sample.

- The probe is a sharp tip and is coupled to the end of a cantilever, which is made of silicon or silicon nitride. When the AFM scans the sample, the cantilever gets deflected as a result of the attractive or repulsive forces between the tip and the sample surface (Fig. 21A). The bending is quantified by a laser beam that reflects on the cantilever back side.

- The forces are finally calculated by combining the information from the laser variation and the known cantilever stiffness. AFM can scan under three different modes depending on the degree of proximity between the probe and the sample, *i.e.* contact, non-contact and tapping mode (also known as intermediate or oscillating mode). The latter is the most common when characterizing NPs. However, it is very sensitive to the free amplitude of the oscillating tip.

- In addition, other parameters, such as tip curvature radius, and surface energy and elasticity of the nanoparticle, influence the final topological values. Nevertheless, these factors can be minimized by plotting the particle height against the free amplitude of the oscillating probe, providing more reliable results. Alternatively, non-contact is preferred when the sample is very sensitive and can be influenced by the tip–sample forces.

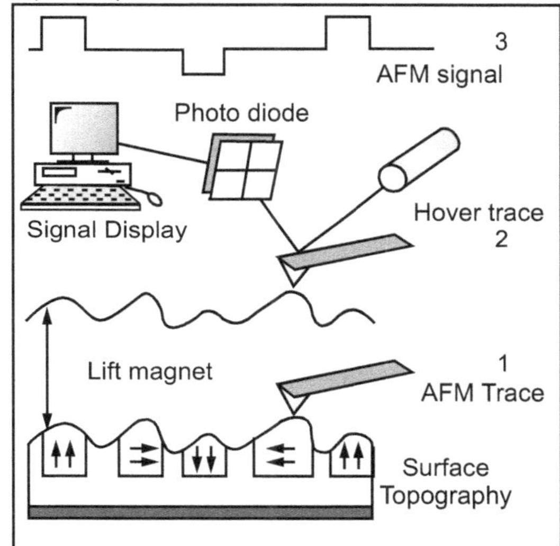


(a)

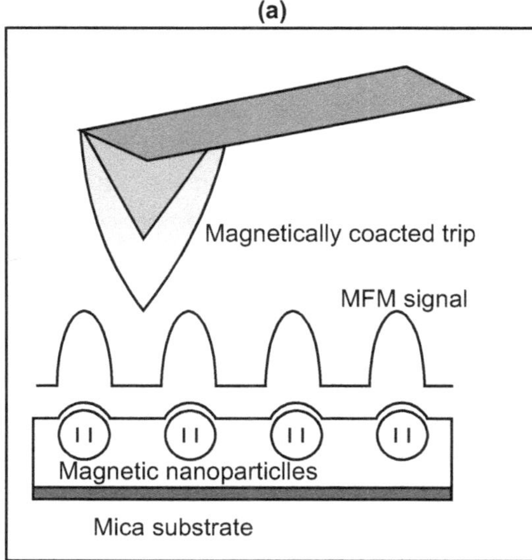


(b)

**Fig. 6.2**

- Fig. 6.2 Schematic of AFM and MFM imaging techniques.

1. An AFM tip scans the surface of a sample to produce a topographical trace,

2. The cantilever is raised to a user-defined height away from the sample surface and the retrace follows the original topographical pattern from the first step;

3. during the retrace, the magnetic signal is scanned and recorded for the sample. In all cases, the signals are recorded *via* the reflection of a laser beam off the back of the cantilever and onto a photodiode, where changes in cantilever deflection are detected.

- In the case of using magnetic force microscopy to scan magnetic NPs on mica substrates, a magnetically coated tip is used to scan the sample surface and an MFM signal is obtained as it interacts magnetically with the sample and its magnetic domains or NPs.

- AFM has the advantage that it does not require any surface modification or coating prior to imaging. Thus, the topological analysis of small NPs (≤6 nm), such as ion-doped Y2O3, has been performed by AFM without any special treatment. Low density materials, which present poor contrast in electron microscopy, have also been characterized.

- For instance, AFM was used to understand the formation mechanism of uniform patchy and hollow rectangular nanoplatelets made of polymer mixtures. Side-by-side comparison between AFM and electron microscopies, i.e. SEM and TEM, showed that AFM provided comparable results when analysing NP sizes. AFM has the advantage that images the sample in three dimensions and allows the characterization of the nanoparticle height.

- Furthermore, it has similar resolution to SEM and TEM, while costing much less and occupying smaller laboratory space. Nevertheless, AFM displays slower scanning times than any electron microscope. Alternatively, spectroscopic techniques, such as DLS and photon correlation spectroscopy (PCS), have also been used to characterize the nanoparticle size. DLS and AFM provided similar results when the sample analysed was monodisperse and uniform.

- However, only AFM could properly characterize NPs with bimodal distribution sizes. The study of alumina nanopowder formation in the solvent by ultrasonic treatment, and posterior sedimentation to a thin film, showed that correlated information could be obtained by PCS and AFM, even though the former analysed liquid samples and the latter solid ones.

- The combination of different technique strengths, such as the high magnification of HRTEM and height measurements of AFM, has helped to understand longstanding problems in nanoscience, like the role of the dendrimer template on the growth of Pt NPs. AFM and XRD were jointly used to characterize Ag NP films, where both techniques provided complementary information.

- In particular, AFM allowed the characterization of the grain size and the nanoparticle coverage of the surface, while XRD identified the preferential growth direction of the particles. Interestingly, at higher NP coverage, AFM showed that the film was made of larger particle grains. Nevertheless, XRD indicated that the crystal size remained the same.

- This apparent contradiction suggested that the larger particles were formed by coalescence of different crystals, yielding larger polycrystalline grains. It is worth mentioning that a densely packed nanoparticle film can be challenging to characterize by AFM, since part of the particles are hidden by their neighbours. Therefore, several algorithms have been developed to estimate the nanoparticle size from the visible part of the image. These algorithms can be applied to densely packed spherical and non-spherical particles.

## 6.5 SEM (SCANNING ELECTRON MICROSCOPY)

- Scanning electron microscopy is a widely used method for the high-resolution imaging of surfaces that can be employed to also characterize nanoscale materials. SEM uses electrons for imaging, much as a light microscope uses visible light. Mazzaglia *et al.* combined *field-emission SEM* (FE-SEM) and XPS measurements to study supramolecular colloidal systems of Au NPs/amphiphilic cyclodextrin.

- These two techniques provided important information on the morphology and nature of the interaction of (thiohexyl carbon chain) $SC_6NH_2$ and (thiohexadecyl carbon chain) $SC_{16}NH_2$ with Au NPs onto the silicon surface. Sinclair and co-workers have reported that SEM and NanoSIMS can be employed to locate Au NPs in cells. SEM analysis illustrated its superiority compared to NanoSIMS for the analysis of inorganic NPs in complex biological systems.

- NanoSIMS has a lower spatial resolution of around 50 nm while SEM is able to achieve resolutions down to 1 nm. The particles tested were Raman-active Au–core NPs and NanoSIMS resulted in somewhat blurred images in certain cases due to its limited resolution. However, NanoSIMS has the unique capability to differentiate between isotopes, although this is not relevant for the case of Au NPs. *High-resolution SEM* (HRSEM) was used by Goldstein *et al.* for the imaging of Au NPs in cells and tissues.

- The straightforward visualization of metallic NPs is assured with this technique, and the sample preparation is fast and easy. However, in case of biological specimens, the need to decrease charging artefacts might make metal coating necessary, thus increasing the risks of radiation damage for the samples. The advantage of HRSEM, compared to other imaging techniques, is the capacity to scale down and study the arrangements of nanometric elements in their wider context.

- It allows the study of the specific spatial arrangement of NPs and thus the examination of possible interactions between them. The results of that study indicated the potential of HRSEM as a relatively simple tool to qualitatively screen the factors that enhance Au NPs penetration, through the skin barrier. It can be considered as a powerful and diverse tool for the study of the interactions between biological systems and metallic nanostructures.

- In another report, SEM and AFM measurements were compared for the same set of NPs, that is, $SiO_2$ and Au NPs on mica or silicon substrates. For example, AFM observations enabled the measurement of the height of a nano-object with sub-nanometric accuracy, but the lateral measurements (along the X and Y axes) had large errors because of the tip/sample convolution.

- In contrast to the AFM, SEM cannot provide any metrological information on the height of the NPs; however, modern SEM can give decent measurements of their lateral dimensions. In fact, the measurements of nearly spherical $SiO_2$ NPs by using both techniques gave similar results, showing the coherency and complementarity of both instruments.

- SEM can be operated in the transmission mode, *i.e.* through the technique called 'transmission *in scanning* electron microscope' (T-SEM). In the transmission mode, advanced NP analysis can be carried out by gaining in-depth information as well as analysis of ensembles of NPs.

- In a paper by Rades et al., the combination of complementary techniques as SEM, T-SEM, EDX and scanning Auger microscopy (SAM) was proven to be a powerful strategy for comprehensive morphological and chemical evaluation of the properties of individual silica and titania NPs.

- On the other hand, methods such as SAXS, DLS, XPS, XRD and BET would be suitable to characterize only the ensembles of the NPs, and not single particles. T-SEM allows a quick examination of the NP shape, though its lateral resolution is limited to NP sizes down to 5–10 nm. TEM provides images with better quality, but T-SEM can be easily combined with EDX for a fast check of the NP size and elemental composition. Hodoroaba *et al.* proved that T-SEM imaging provides a size distribution that is slightly broader than that obtained by TEM.

- For small $SiO_2$ NPs, the precise delimitation of the particles in the T-SEM mode is definitely constrained by the lower spatial resolution achieved compared to that of conventional TEM. In addition, with the T-SEM, the surface layer of the particles might not be always easily detected. The same author noted in another paper that the conventional SEM imaging mode could not detect the NPs on the back side of the support film that was required for the observations. Therefore, an explicit knowledge of the T-SEM operator is needed for the measurements.

- The authors observed that the obtained $SiO_2$ NP size distributions by SEM and TSEM in their work and for various conditions agreed well with each other, within the associated measurement uncertainties. In another report, 3D reconstruction by Focused Ion Beam (FIB) cutting and SEM imaging were combined to comprehend the evolution of pore volume, pore shape and other parameters during the two-step sintering of ZnO NPs.

- In this way, the sintering process at the nanoscale for such particles can be better understood. Ni- and Cu-co-doped zinc oxide NPs prepared by the co-precipitation method were investigated by Ashokkumar and Muthukumaran by microstructure, optical and FTIR measurements. The depicted shape by SEM was in good agreement with the mathematical determinations from XRD, whereas FTIR provided important information on chemical bonding.

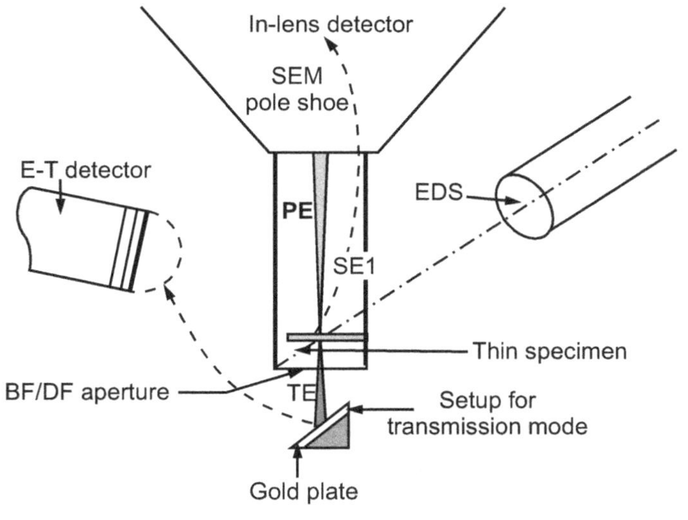


**Fig. 6.3**

Fig. 6.3 Scheme of a SEM/EDS system operating in the transmission mode with the Zeiss single-unit transmission setup (PE: primary electrons; SE1: secondary electrons emitted at the point of impact of the PE on the sample; TE: transmitted electrons; BF: bright field; DF: dark field; E–T: Everhart–Thornley detector).

## 6.6 TEM (TRANSMISSION ELECTRON MICROSCOPY)

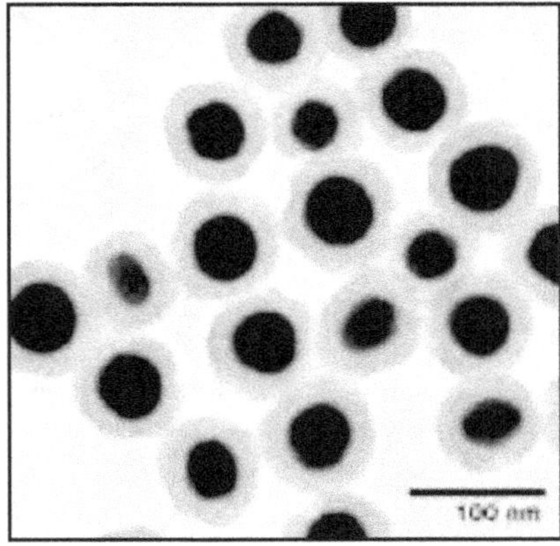


**Fig. 6.4**

- Transmission electron microscopy (TEM) is a high magnification measurement technique that images the transmission of a beam of electrons through a sample. Amplitude and phase variations in the transmitted beam provide imaging contrast that is a function of the sample thickness (the amount of material that the electron beam must pass through) and the sample material (heavier atoms scatter more electrons and therefore have a smaller electron mean free path than lighter atoms). Because this technique uses electrons rather than light to illuminate the sample, TEM imaging has significantly higher resolution than light-based imaging techniques.

- Successful imaging of nanoparticles using TEM depends on the contrast of the sample relative to the background. Samples are prepared for imaging by drying nanoparticles on a copper grid that is coated with a thin layer of carbon. Materials with electron densities that are significantly higher than amorphous carbon are easily imaged.

- These materials include most metals (e.g., silver, gold, copper, aluminum), most oxides (e.g., silica, aluminum oxide, titanium oxide), and other particles such as polymer nanoparticles, carbon nanotubes, quantum dots, and magnetic nanoparticles. TEM imaging is the preferred method to directly measure the particle size, grain size, size distribution, and morphology of nanoparticles. Sizing accuracy is typically within 3% of the actual value.

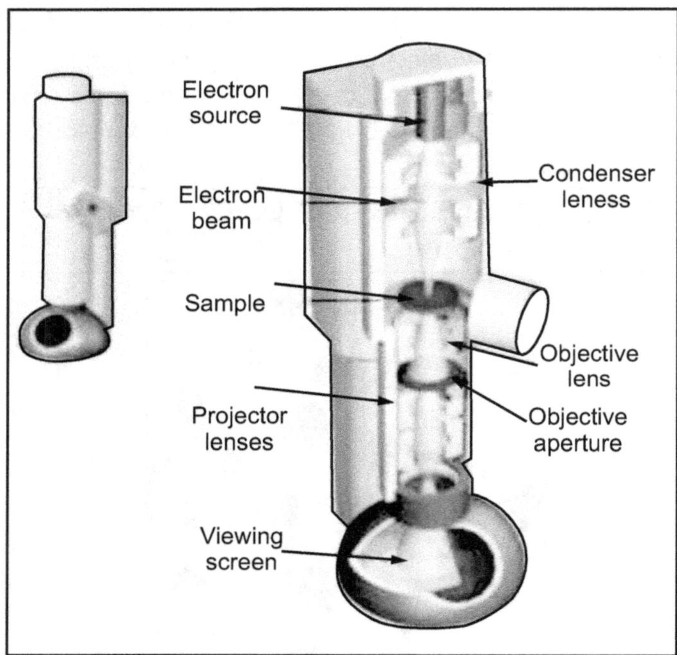


**Fig. 6.5**

- At nanoComposix, we use a JEOL 1010 transmission electron microscope operating at an accelerating voltage of 100 keV and an AMT XR41-B 4-megapixel (2048 x 2048) bottom mount CCD camera. The camera's finite-conjugate optical coupler provides high resolution and flat focus with less than 0.1% distortion for magnifications as high as 150,000x.

## 6.7 EDAX APPLICATIONS AND INTERPRETATION OF RESULTS

- Energy-dispersive X-ray spectroscopy (EDS, EDX, EDXS or XEDS), sometimes called energy dispersive X-ray analysis (EDXA) or energy dispersive X-ray microanalysis (EDXMA), is an analytical technique used for the elemental analysis or chemical characterization of a sample.

- It relies on an interaction of some source of X-ray excitation and a sample. Its characterization capabilities are due in large part to the fundamental principle that each element has a unique atomic structure allowing a unique set of peaks on its electromagnetic emission spectrum (which is the main principle of spectroscopy).

- To stimulate the emission of characteristic X-rays from a specimen, a high-energy beam of charged particles such as electrons or protons (see PIXE), or a beam of X-rays, is focused into the sample being studied. At rest, an atom within the sample contains ground state (or unexcited) electrons in discrete energy levels or electron shells bound to the nucleus.

- The incident beam may excite an electron in an inner shell, ejecting it from the shell while creating an electron hole where the electron was. An electron from an outer, higher-energy shell then fills the hole, and the difference in energy between the higher-energy shell and the lower energy shell may be released in the form of an X-ray.

- The number and energy of the X-rays emitted from a specimen can be measured by an energy-dispersive spectrometer. As the energies of the X-rays are characteristic of the difference in energy between the two shells and of the atomic structure of the emitting element, EDS allows the elemental composition of the specimen to be measured.

- **Four Primary Components of the EDS Setup are**
  1. The excitation source (electron beam or x-ray beam)
  2. The X-ray detector
  3. The pulse processor
  4. The analyzer.

- Electron beam excitation is used in electron microscopes, scanning electron microscopes (SEM) and scanning transmission electron microscopes (STEM). X-ray beam excitation is used in X-ray fluorescence (XRF) spectrometers.

- A detector is used to convert X-ray energy into voltage signals; this information is sent to a pulse processor, which measures the signals and passes them onto an analyzer for data display and analysis. The most common detector used to be Si(Li) detector cooled to cryogenic temperatures with liquid nitrogen. Now, newer systems are often equipped with silicon drift detectors (SDD) with Peltier cooling systems.

## 6.8 EMERGING NANO MATERIAL

- Graphene, a typical two-dimensional material (2DM), which presents sheet-like faveolated structure, has attracted an extraordinary amount of interest in both basic and applied researches. Many extraordinary properties, such as high surface area, ultrahigh electron mobility and excellent thermal conductivity, make graphene successfully used in the fabrication of hybrid materials, optical devices, chemical/biological sensors, lithium ion batteries and supercapacitors.

- The great success of graphene throws new light on discovering more 2D layered nanomaterials composed of atomic sheets. In recent ten years, more than hundreds of graphene-like nanomaterials have been synthesized, such as transition metal nanomaterials, graphitic carbon nitride, hexagonal boron nitride, black phosphorus and some organic polymers, greatly enriching the family of 2DMs.

- Compared with a single element of graphene, graphene-like 2DMs composed of multiple elements that possess more versatility, greater flexibility and better functionality. Featuring ultrathin thickness and 2D morphology structure, these graphene-like nanomaterials present unique physicochemical and electronic properties compared to their bulk congeries.

- On account of their outstanding characters and large specific surface areas, these graphene-like 2D layered nanomaterials give rise to various potential applications, such as electronic devices, catalysis, sensors and energy storage. In addition, they hold excellent biocompatibility and can directly disperse in aqueous solutions without modification, making them very suitable for biosensing applications.

- Some graphene-like 2DMs can directly interact with biomolecules, forming surface modified 2DMs. More importantly, a wide range of protocols in the fabrication of 2D nanosheets offers great opportunities for generating complex functional nanostructures, which optimize their properties and achieve remarkable performances for diversified applications.

- Nanotechnologies are attracting increasing investments from both governments and industries around the world, which offers great opportunities to explore the new emerging nanodevices, such as the Carbon Nanotube and Nanosensors. This technique exploits the specific properties which arise from structure at a scale characterized by the interplay of classical physics and quantum mechanics.

- It is difficult to predict these properties a priori according to traditional technologies. Nanotechnologies will be one of the next promising trends after MOS technologies. However, there has been much hype around nanotechnology, both by those who want to promote it and those who have fears about its potentials. This paper gives a deep survey regarding different aspects of the new nanotechnologies, such as materials, physics, and semiconductors respectively, followed by an introduction of several state-of-the-art nanodevices and then new nanotechnology features.

- Since little research has been carried out on the toxicity of manufactured nanoparticles and nanotubes, this paper also discusses several problems in the nanotechnology area and gives constructive suggestions and predictions.

- Nanotechnology, introduced almost half a century ago, is an active research area with both novel science and useful applications that has gradually established itself in the past two decades. Nanotechnology – a term encompassing the science, engineering, and applications of submicron materials – involves the harnessing of the unique physical, chemical, and biological properties of nanoscale substances in fundamentally new and useful ways.

- The economic and societal promise of nanotechnology has led to investments by governments and companies around the world. The complexities and intricacies of nanotechnology, still in the early stage of development, and the broad scope of these potential applications, have become increasingly important.

- A nanometer is one-billionth of a meter. For example, a sheet of paper is about 100,000 nanometers thick; a single gold atom is about a third of a nanometer in diameter. Dimensions between approximately 1 and 100 nanometers are known as the nanoscale. Nanotechnology is the understanding and control of matter at the dimensions between approximately 1 and 100 nanometers, where unique phenomena enable novel applications.

- Encompassing nanoscale science, engineering and technology, nanotechnology involves imaging, measuring, modeling, and manipulating matter at this length scale. The transformative and general purpose prospects associated with nanotechnology have stimulated more than 60 countries to invest in national nanotechnology research and development programs.

- Unusual physical, chemical, and biological properties can emerge in materials at the nanoscale. These properties may appear dramatically different in important ways from the properties of bulk materials and single atoms or molecules.

- Using structures designed at this extremely small scale, there exist opportunities to build materials, devices, and systems with nano-properties that can not only enhance existing technologies but also offer novel features with potentially far-reaching technical, economic, and societal implications . Nanotechnology products can be used for the design and processes in various areas.

- It has been demonstrated that nanotechnology has many unique characteristics, and can significantly fix the current problems which the non-nanotechnology faced, and may change the requirement and organization of design processes with its unique features. Nanotechnology deals with the production and applications at scales ranging from a few nanometers to submicron dimensions, as well as the integration of the resulting nanostructures into larger systems.

- It also involves the investigation of individual atoms. Particularly, the conventional analytic aspects of nanotechnology must yield a certain synthetic approach, which is similar with non-nanotechnology. This action will be conducive to the creation of new functions exhibited by nanoscale structural units through their mutual interactions, even though these functionalities are not properties of the isolated units.

- We can use the term nanoarchitectonics to express this innovation of nanotechnology : it is a technology system aimed at arranging nanoscale structural units, a group of atoms, molecules, or nanoscale functional components, into a configuration that creates a novel functionality through mutual interactions among those units.

- **These Very Small Structures in Nanoscale are Intensely Interesting for Many Reasons :**

  1. Many properties mystify us. For example, how do electrons move through organometallic nanowires?

  2. They are challenging to make. For example, synthesizing or fabricating ordered arrays and patterns of nanoscale units poses fascinating problems.

  3. Studying these structures leads to new phenomena, since many nanoscale structures have been inaccessible and/or off the beaten scientific track.

  4. Nanostructures are in a range of sizes in which quantum phenomena, especially quantum entanglement and other reflections of the wave character of matter, would be expected to be important.

  5. The nanometer-sized, functional structures that carry out many of the most sophisticated tasks open up an exciting frontier of biology.

- The nanomaterials field includes subfields which develop or study materials having unique properties arising from their nanoscale dimensions.

- Interface and colloid science has given rise to many materials which may be useful in nanotechnology, such as carbon nanotubes and other fullerenes, and various nanoparticles and nanorods. Nanomaterials with fast ion transport are related also to nanoionics and nanoelectronics.

- Nanoscale materials can also be used for bulk applications; most present commercial applications of nanotechnology are of this flavor.

- Progress has been made in using these materials for medical applications; see Nanomedicine.

- Nanoscale materials such as nanopillars are sometimes used in solar cells which combats the cost of traditional silicon solar cells.

- Development of applications incorporating semiconductor nanoparticles to be used in the next generation of products, such as display technology, lighting, solar cells and biological imaging; see quantum dots.

- Recent application of nanomaterials include a range of biomedical applications, such as tissue engineering, drug delivery, and biosensors.

## 6.9 NANO TUBES

- Carbon nanotubes (CNTs) are cylindrical molecules that consist of rolled-up sheets of single-layer carbon atoms (graphene). They can be single-walled (SWCNT) with a diameter of less than 1 nanometer (nm) or multi-walled (MWCNT), consisting of several concentrically interlinked nanotubes, with diameters reaching more than 100 nm. Their length can reach several micrometers or even millimeters.

- Like their building block graphene, CNTs are chemically bonded with sp2 bonds, an extremely strong form of molecular interaction. This feature combined with carbon nanotubes' natural inclination to rope together via van der Waals forces, provide the opportunity to develop ultra-high strength, low-weight materials that possess highly conductive electrical and thermal properties. This makes them highly attractive for numerous applications.

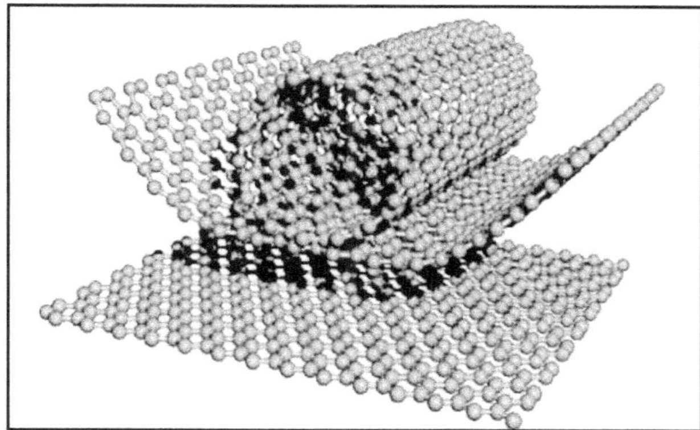

**Fig. 6.6**

- Schematic of how graphene could roll up to form a carbon nanotube.

- The rolling-up direction (rolling-up or chiral vector) of the graphene layers determines the electrical properties of the nanotubes. Chirality describes the angle of the nanotube's hexagonal carbon-atom lattice.

- Armchair nanotubes – so called because of the armchair-like shape of their edges – have identical chiral indices and are highly desired for their perfect conductivity. They are unlike zigzag nanotubes, which may be semiconductors. Turning a graphene sheet a mere 30 degrees will change the nanotube it forms from armchair to zigzag or vice versa.

- While MWCNTs are always conducting and achieve at least the same level of conductivity as metals, SWCNTs' conductivity depends on their chiral vector: they can behave like a metal and be electrically conducting; display the properties of a semi-conductor; or be non-conducting. For example, a slight change in the pitch of the helicity can transform the tube from a metal into a large-gap semiconductor.

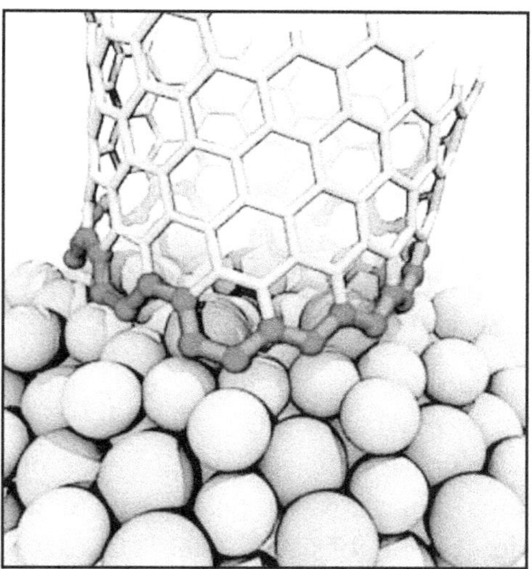

**Fig. 6.7**

This illustration shows the interface between a growing carbon nanotube and a cobalt-tungsten catalyst. The atomic arrangement of the catalyst forces the nanotube to quickly transition from zigzag (blue) to armchair (red), which ultimately grows a nanotube . (Image: Evgeni Penev/Rice University)

Apart from their electrical properties, which they inherit from graphene, CNTs also have unique thermal and mechanical properties that make them intriguing for the development of new materials:

- Their mechanical tensile strength can be 400 times that of steel;

- They are very light-weight – their density is one sixth of that of steel;

- Their thermal conductivity is better than that of diamond;

- They have a very high aspect ratio greater than 1000, i.e. in relation to their length they are extremely thin;

- A tip-surface area near the theoretical limit (the smaller the tip-surface area, the more concentrated the electric field, and the greater the field enhancement factor);

- Just like graphite, they are highly chemically stable and resist virtually any chemical impact unless they are simultaneously exposed to high temperatures and oxygen - a property that makes them extremely resistant to corrosion;

- Their hollow interior can be filled with various nanomaterials, separating and shielding them from the surrounding environment - a property that is extremely useful for nanomedicine applications like drug delivery.

All these properties make carbon nanotubes ideal candidates for electronic devices, chemical/electrochemical and biosensors, transistors, electron field emitters, lithium-ion batteries, white light sources, hydrogen storage cells, cathode ray tubes (CRTs), electrostatic discharge (ESD) and electrical-shielding applications.

## 6.10 NANO RODS AND OTHER NANO STRUCTURES

- GaGdN nanorod structures were grown on Si(001) substrates with native SiO2 layer by radio-frequency plasma-enhanced MBE (RF-MBE) under nitrogen-rich conditions. After thermal cleaning of Si substrates at 870°C for 15 min, LT GaN buffer layers were grown at 450°C for 60 s.

- Then annealing was performed at 850°C for 10 min in the plasma-activated nitrogen atmosphere followed by the growth of GaN nanorods at 800°C for 15 min. Then GaGdN nanorods were grown. Finally GaN layers were deposited to prevent the oxidation of GaGdN nanorods' surface.

- When GaGdN nanorods were grown at high substrate temperatures, for example at 850°C, the nanorod diameter is kept constant independent of nanorod length.

- However, when grown at LTs, for example at 550°C, to increase the incorporated Gd concentration, the use of high Gd flux gradually increased the diameter of GaGdN nanorods with the progress of growth and the crystalline quality of nanorods degraded. To solve this problem, the GaGdN/GaN SL structure nanorods were grown.

- Fig. 6.8 (a) Schematic view of grown GaGdN/GaN SL nanorod structures. Fig. 6.8 (b) Cross-sectional SEM images for GaGdN/GaN SL nanorod structures as a function of Gd beam equivalent pressure (BEP).

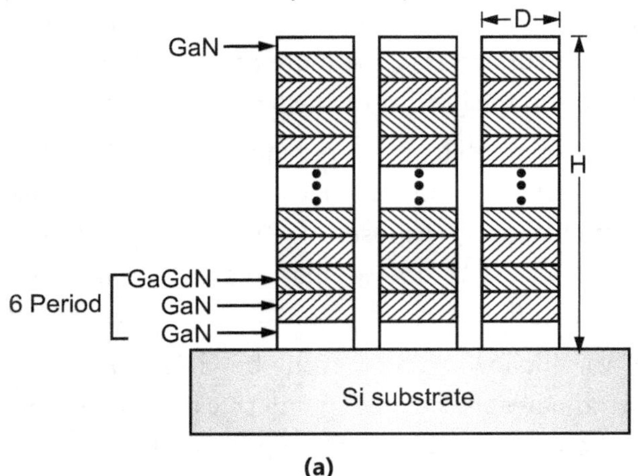


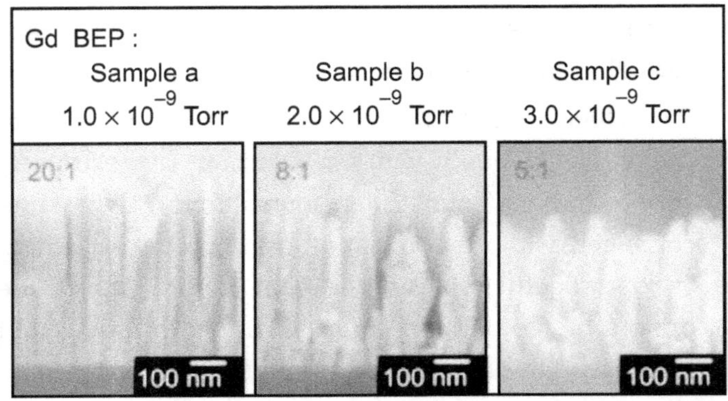


(b)

**Fig. 6.8**

- Cross-sectional scanning electron microscopy (SEM) photographs for these GaGdN/GaN SL structure nanorods with different nanorod aspect ratio (height-to-diameter ratio). On these GaGdN nanorods, clear hysteresis and saturation were observed in the M-H curves at RT.

- Furthermore, by changing the nanorod aspect ratio, the vertical magnetization was improved. To realize the CP-LDs, vertically magnetized DMS layers are essentially required.

### ZnO Nanorods

- Nanorods or nanowires are considered as functional components in electronics and optoelectronics (Dekker, 1999; Hu et al., 1999, 2001). The synthesis of such nanostructures with embedded quantum structures can enable new electrical or optical properties due to the formation of quantum-confined states.

- ZnO is an interesting semiconducting oxide due to its high exciton binding energy, ability to grow as nanowires, and light optical emission in the visible range. STEM is an ideal tool to experimentally study the correlations between light optical emission profiles through CL and the atomic structure of ZnO nanostructures.

- For the case study reviewed in the following, ZnO nanorods were grown using catalyst-free metal-organic vapor-phase epitaxy. The growth of quasi-1D nanostructures, that is, nanorods, was facilitated by the preferential growth along the c-axis of ZnO. An Mg precursor was introduced into the growth chamber to epitaxially grow Zn0.8Mg0.2O layers on the tip of the nanorods.

- In the absence of catalyst particles, this growth mechanism results into minimal intermixing, sharp interfaces with ZnO, and tunable well thicknesses.

Details about the growth process are reported elsewhere. Precise thickness control enables the generation of quantum-confined states in the bandgap and, thus, an increasing blue shift with decreasing layer thickness.

- Fig. 6.9 shows electron micrographs of the nanorods at various magnifications. Scanning electron microscopy revealed that the nanorods were well aligned with respect to each other with homogeneous lengths and diameters. ADF STEM images recorded with increasing magnifications clearly reveal the atomic structure of the quantum-wells. The interfaces between ZnO and the Zn0.8Mg0.2O layers appear atomically sharp.

- The Zn0.8Mg0.2O layers have somewhat darker contrast than ZnO due to the Z-contrast mechanism. The atomic resolution image of the quantum-well structure reveals the presence of a superlattice in the growth direction, which possibly indicates structural ordering of oxygen vacancies.

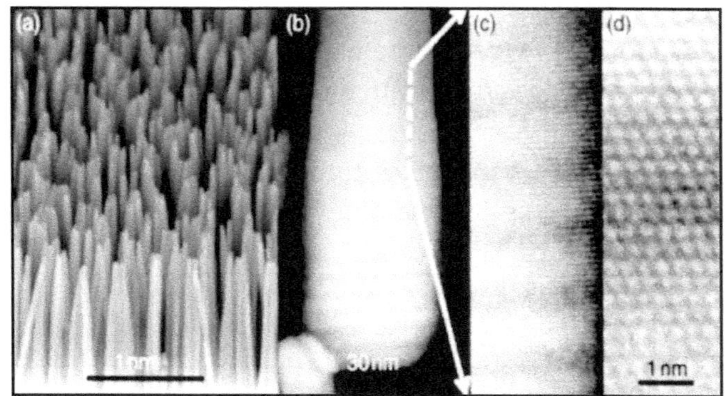


**Fig. 6.9**

Fig. 6.9 Scanning electron microscopy (SEM) image showing a vertically aligned array of ZnO nanorods, with $Zn_{0.8}Mg_{0.2}O$/ZnO multiple quantum-wells at the tip. (b)–(d) HAADF images with increasing magnification showing the quantum-well structure.

## 6.11 LB TECHNIQUE

- A Langmuir–Blodgett (LB) film is a nanostructured system formed when Langmuir films or Langmuir monolayers (LM) are transferred from the liquid-gas interface to solid supports during the vertical passage of the support through the monolayers. LB films can contain one or more monolayers of an organic material, deposited from the surface of a liquid onto a solid by immersing (or emersing) the solid substrate into (or from) the liquid.

- A monolayer is adsorbed homogeneously with each immersion or emersion step, thus films with very accurate thickness can be formed. This thickness is accurate because the thickness of each monolayer is known and can therefore be added to find the total thickness of a Langmuir–Blodgett film.

- The monolayers are assembled vertically and are usually composed either of amphiphilic molecules (see Chemical polarity) with a hydrophilic head and a hydrophobic tail (example: fatty acids) or nowadays commonly of nanoparticles.

- Langmuir–Blodgett films are named after Irving Langmuir and Katharine B. Blodgett, who invented this technique while working in Research and Development for General Electric Co.

- Langmuir films are formed when amphiphilic (surfactants) molecules or nanoparticles are spread on the air at an air–water interface. Surfactants (or surface-acting agents) are molecules with hydrophobic 'tails' and hydrophilic 'heads'. When surfactant concentration is less than the minimum surface concentration of collapse and it is completely insoluble in water, the surfactant molecules arrange themselves as shown in Fig. 1.10 below.

- This tendency can be explained by surface-energy considerations. Since the tails are hydrophobic, their exposure to air is favoured over that to water. Similarly, since the heads are hydrophilic, the head–water interaction is more favourable than air–water interaction. The overall effect is reduction in the surface energy (or equivalently, surface tension of water).

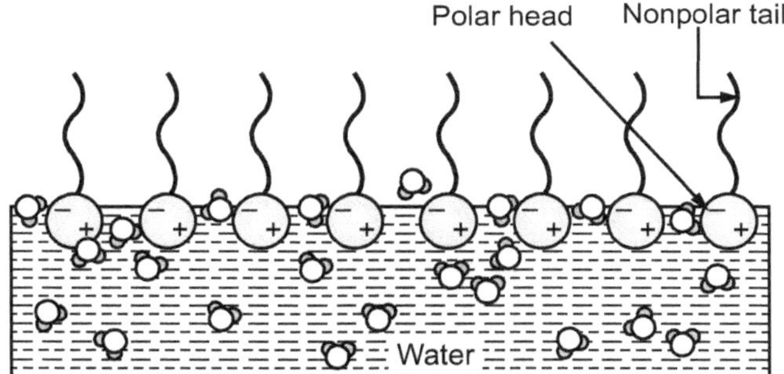


**Fig. 6.10: Surfactant molecules arranged on an air–water interface**

- For very small concentrations, far from the surface density compatible with the collapse of the monolayer (which leads to polylayers structures) the surfactant molecules execute a random motion on the water–air interface. This motion can be thought to be similar to the motion of ideal-gas molecules enclosed in a container. The corresponding thermodynamic variables for the surfactant system are, surface pressure ($\displaystyle \Pi$), surface area (A) and number of surfactant molecules (N). This system behaves similar to a gas in a container.

- The density of surfactant molecules as well as the surface pressure increases upon reducing the surface area A ('compression' of the 'gas'). Further compression of the surfactant molecules on the surface shows behavior similar to phase transitions.

- The 'gas' gets compressed into 'liquid' and ultimately into a perfectly closed packed array of the surfactant molecules on the surface corresponding to a 'solid' state. The liquid state is usually separated in the liquid-expanded and liquid-condensed states. All the Langmuir film states are classified according to the compressionality factor of the films, defined as -A(d ({\displaystyle \Pi })/dA), usually related to the in-plane elasticity of the monolayer.

- The condensed Langmuir films (in surface pressures usually higher than 15 mN/m - typically 30 mN/m) can be subsequently transferred onto a solid substrate to create highly organized thin film coatings. Langmuir–Blodgett troughs

- Besides LB film from surfactants depicted in Fig. 6.10, similar monolayers can also be made from inorganic nanoparticles.

- Many possible applications have been suggested over years for LM and LB films. Their characteristics are extremely thin films and high degree of structural order. These films have different optical, electrical and biological properties which are composed of some specific organic compounds.

- Organic compounds usually have more positive responses than inorganic materials for outside factors (pressure, temperature or gas change). LM films can be used also as models for half a cellular membrane.

- LB films consisting of nanoparticles can be used for example to create functional coatings, sophisticated sensor surfaces and to coat silicon wafers.

- LB films can be used as passive layers in MIS (metal-insulator-semiconductor) which have more open structure than silicon oxide, and they allow gases to penetrate to the interface more effectively.

- LB films also can be used as biological membranes. Lipid molecules with the fatty acid moiety of long carbon chains attached to a polar group have received extended attention because of being naturally suited to the Langmuir method of film production.

- This type of biological membrane can be used to investigate: the modes of drug action, the permeability of biologically active molecules, and the chain reactions of biological systems.

- Also, it is possible to propose field effect devices for observing the immunological response and enzyme-substrate reactions by collecting biological molecules such as antibodies and enzymes in insulating LB films.

- Anti-reflective glass can be produced with successive layers of fluorinated organic film.

- The glucose biosensor can be made of poly(3-hexyl thiopene) as Langmuir–Blodgett film, which entraps glucose-oxide and transfers it to a coated indium-tin-oxide glass plate.

- UV resists can be made of poly(N-alkylmethacrylamides) Langmuir–Blodgett film.

- UV light and conductivity of a Langmuir–Blodgett film.

- Langmuir–Blodgett films are inherently 2D-structures and can be built up layer by layer, by dipping hydrophobic or hydrophilic substrates into a liquid sub-phase.

- Langmuir–Blodgett patterning is a new paradigm for large-area patterning with mesostructured features.

- Recently, it has been demonstrated that Langmuir-Blodgett is an effective technique even to produce ultra-thin films of emerging two-dimensional layered materials on a large scale.

## 6.12 SOFT LITHOGRAPHY MICROWAVE ASSISTED SYNTHESIS

- In technology, soft lithography is a family of techniques for fabricating or replicating structures using "elastomeric stamps, molds, and conformable photomasks". It is called "soft" because it uses elastomeric materials, most notably PDMS.

- Soft lithography is generally used to construct features measured on the micrometer to nanometer scale. According to Rogers and Nuzzo (2005), development of soft lithography expanded rapidly from 1995 to 2005. Soft lithography tools are now commercially available.

- Soft lithography has some unique advantages over other forms of lithography (such as photolithography and electron beam lithography).

They include the following:

- ➤ Lower cost than traditional photolithography in mass production

- ➤ Well-suited for applications in biotechnology

- ➤ Well-suited for applications in plastic electronics

➤ Well-suited for applications involving large or nonplanar (nonflat) surfaces

- More pattern-transferring methods than traditional lithography techniques (more "ink" options)

- Does not need a photo-reactive surface to create a nanostructure

- Smaller details than photolithography in laboratory settings (~30 nm vs ~100 nm). The resolution depends on the mask used and can reach 6 nm.

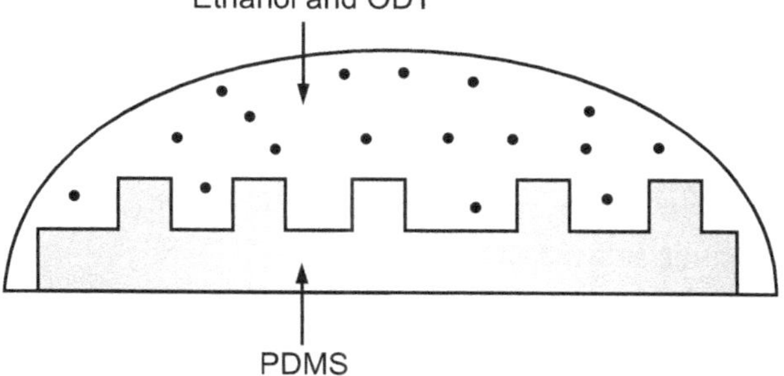


**Fig. 6.11 : "Inking" a stamp. PDMS stamp with pattern is placed in Ethanol and ODT**

- Nanolithography is a growing field of techniques within nanotechnology dealing with the engineering (etching, writing, printing) of nanometer-scale structures. From Greek, the word can be broken up into three parts: "nano" meaning dwarf, "lith" meaning stone, and "graphy" meaning to write, or "tiny writing onto stone."

- Today, the word has evolved to cover the design of structures in the range of $10^{-9}$ to $10^{-6}$ meters, or structures in the nanometer range. Essentially, field is a derivative of lithography, only covering significantly smaller structures. All nanolithographic techniques can be separated into two categories: those that etch away molecules leaving behind the desired structure, and those that directly write the desired structure to a surface (similar to the way a 3D printer creates a structure).

- The field of nanolithography was born out of the need to increase the number of transistors in an integrated circuit in order to maintain Moore's Law. While lithographic techniques have been around since the late 18th century, none were applied to nanoscale structures until the mid-1950s. With evolution of the semiconductor industry, demand for techniques capable of producing micro- and nano-scale structures skyrocketed.

- Photolithography was applied to these structures for the first time in 1958 beginning the age of nanolithography. Since then, photolithography has become the most commercially successful technique, capable of producing sub-100 nm patterns.

- There are several techniques associated with the field, each designed to serve its many uses in the medical and semiconductor industries. Breakthroughs in this field contribute significantly to the advancement of nanotechnology, and are increasingly important today as demand for smaller and smaller computer chips increases. Further areas of research deal with physical limitations of the field, energy harvesting, and photonics.

**Optical Lithography**

- Optical Lithography (or photolithography) is one of the most important and prevalent sets of techniques in the nanolithography field. Optical lithography contains several important derivative techniques, all that use very short light wavelengths in order to change the solubility of certain molecules, causing them to wash away in solution, leaving behind a desired structure.

- Several optical lithography techniques require the use of liquid immersion and a host of resolution enhancement technologies like phase-shift masks (PSM) and optical proximity correction (OPC). Some of the included techniques in this set include multiphoton lithography, X-Ray lithography, light coupling nanolithography (LCM), and extreme ultraviolet lithography (EUVL). This last technique is considered to be the most important next generation lithography (NGL) technique due to its ability to produce structures accurately down below 30 nanometers.

**Electron-Beam Lithography**

- Electron beam lithography (EBL) or electron-beam direct-write lithography (EBDW) scans a focused beam of electrons on a surface covered with an electron-sensitive film or resist (e.g. PMMA or HSQ) to draw custom shapes. By changing the solubility of the resist and subsequent selective removal of material by immersion in a solvent, sub-10 nm resolutions have been achieved.

- This form of direct-write, maskless lithography has high resolution and low throughput, limiting single-column e-beams to photomask fabrication, low-volume production of semiconductor devices, and research&development. Multiple-electron beam approaches have as a goal an increase of throughput for semiconductor mass-production.

- EBL can be utilized for selective protein nanopatterning on a solid substrate, aimed for ultrasensitive sensing.

## Scanning Probe Lithography

- Scanning probe lithography (SPL) is another set of techniques for patterning at the nanometer-scale down to individual atoms using scanning probes, either by etching away unwanted material, or by directly-writing new material onto a substrate.

- Some of the important techniques in this category include dip-pen nanolithography, thermochemical nanolithography, thermal scanning probe lithography, and local oxidation nanolithography. Dip-pen nanolithography is the most widely used of these techniques.

## Nanoimprint Lithography

- Nanoimprint lithography (NIL), and its variants, such as Step-and-Flash Imprint Lithography and laser assisted directed imprint (LADI) are promising nanopattern replication technologies where patterns are created by mechanical deformation of imprint resists, typically monomer or polymer formations that are cured by heat or UV light during imprinting.[citation needed] This technique can be combined with contact printing and cold welding. Nanoimprint lithography is capable of producing patterns at sub-10 nm levels.

## Magnetolithography

- (ML) is based on applying a magnetic field on the substrate using paramagnetic metal masks call "magnetic mask". Magnetic mask which is analog to photomask define the spatial distribution and shape of the applied magnetic field.

- The second component is ferromagnetic nanoparticles (analog to the Photoresist) that are assembled onto the substrate according to the field induced by the magnetic mask.

## Nanosphere Lithography

- This technique uses self-assembled monolayers of spheres (typically made of polystyrene) as evaporation masks. This method has been used to fabricate arrays of gold nanodots with precisely controlled spacings.

## Proton Beam Writing

- This technique uses a focused beam of high energy (MeV) protons to pattern resist material at nanodimensions and has been shown to be capable of producing high-resolution patterning well below the 100 nm mark.

## Charged-Particle Lithography

- This set of techniques include ion- and electron-projection lithographies. Ion beam lithography uses a focused or broad beam of energetic lightweight ions (like He+) for transferring pattern to a surface. Using Ion Beam Proximity Lithography (IBL) nano-scale features can be transferred on non-planar surfaces.

## 6.13 SELF-ASSEMBLY

- Self-assembly refers to an autonomous process whereby disordered building blocks gradually form larger, well-organized patterns, driven by mutual interactions of the building blocks toward reducing the system free energy. The process obeys physical laws and can occur at all scales, in the form of static and dynamic self-assembly.

- Both static and dynamic self-assemblies form stable or meta-stable patterns with highly ordered structures. Static self-assembly is commonly a relatively slow process and formed from the simple aging process, with the formed patterns integrally stable via a free energy minimization process. Dynamic self-assembly commonly forms stable patterns that are not in equilibrium.

- As a simple and low-cost approach, the bottom-up self-assembly strategy shows unique merits in large scale batch fabrication of ordered patterns, compared with top down nano fabrication techniques, which usually require expensive cleanroom facilities. The self-assembly of micro / nanoparticles, among various kinds of building blocks, has generated intense interest because the self-assembled patterns commonly possess unique physical properties and find various applications in the fields of nanophotonics, solar cells, catalysts, data storage, and so on.

- For instance, statically self-assembled patterns with micro / nano nanoparticles show features with a comparable size to the wavelength of incident light, inducing selective Bragg's diffraction of light and interference effects. While the light wavelength is within the range of visible light, the material will show a particular color.

- The color can even change with one's viewing orientation due to the alternation of the constructive and destructive interference. Distinguished from pigmentary color, this kind of color is generated from periodical micro / nano structures. Various brilliant colors in nature like the feathers of birds, shells of beetles, the skin of chameleons, and petals of flowers originate from the surface imprinted periodic micro / nano patterns, which offer interminable sources and examples for researchers to develop photonic materials with excellent optical properties mimicking nature, and even going beyond nature.

- Unlike static self-assembly, the dynamic system is forcefully prevented from reaching a free energy minimum and requires a continuous supply of external energy to balance the intrinsic interactions between building blocks, sustaining stable dynamic patterns with on-demand controllable configurations. Short-range forces either attractive or repulsive such as electrostatic interaction, van der Waals attraction and steric repulsion, all work during the dynamic self-assembly process, making it more complicated than static cases.

- It is sensitive to tiny and local changes, disturbances among the building blocks, and the supplied external energy. Dynamic self-assembly is very common in nature and the biological world, such as starling flocks, schooling fish, and bacterial swarms; however, the development of artificial dynamic self-assembly lags behind the static systems. The dynamic assembled pattern shows its merit in the adaptability compared with the static ones.

- To date, researchers have developed some dynamic systems with various kinds of external energy input, such as magnetic field, electric field, ultrasound field, light field, and hybrid energy source, which provide promising means to realize robot swarms at the small scales. Recently, with a combination of magnetic assembly and controlled locomotion, microrobotic swarms of magnetic colloid particles with reconfigurable patterns have been reported, resulting in various applications in cargo transportation and delivery, magnetic hyperthermia, anti-diffusion, and heavy metal removal.

- The dynamic patterns are generated and controlled on-demand, originating from dipole–dipole magnetic interactions between building blocks and fluidic drag effects. A magnetic field is used as the external energy source due to its merits in long-range and precise actuation. Moreover, a low-frequency magnetic field can penetrate deep tissues and is harmless to

biological organisms, facilitating future in vivo applications of the dynamic swarming colloid particles, such as drug delivery and thrombolysis.

- The image on the cover of this issue of *Materials Today* shows a large-scale and multi-layer photonic crystal with periodically distributed defects obtained from the static self-assembly process of a highly concentrated superhydrophilic mesoporous silica nanospheres on a silicon wafer via evaporation. The self-assembled vortex-like pattern is built from the uniform mesoporous silica particles with a size of approximate 100?nm.

- A lot of line defects were formed during the evaporation of the solvent. The line defects show well-organized arrangement on the silica pattern and reach up to about 2?mm in length originating from the center of the pattern. This kind of photonic crystal, offering plenty of defects with uniform and periodical arrangements, may provide potential applications in nano photonics, environmental sensing, and anti-fouling materials.

## SUMMARY

- Nanoscience and nanotechnology are still undergoing constant growth, and the scientific community is rather aware that there may be certain differences between the way analytical characterization methods operate for nanomaterials, in comparison with their more 'traditional' modes of use for more 'conventional' (macroscopic) materials.

- FTIR is one techniques that help in the determination of the structure, composition, size and other basic features of the NPs. Fourier transform infrared spectroscopy (FTIR) is a technique based on the measurement of the absorption of electromagnetic radiation with wavelengths within the mid-infrared region ($4000–400 \text{ cm}^{-1}$).

- X-ray diffraction (XRD) is one of the most extensively used techniques for the characterization of NPs. Typically, XRD provides information regarding the crystalline structure, nature of the phase, lattice parameters and crystalline grain size

- AFM (Atomic force microscopy) is a microscopy technique capable of creating three-dimensional images of surfaces at high magnification.

- SEM (Scanning electron microscopy) is a widely used method for the high-resolution imaging of surfaces that can be employed to also characterize nanoscale materials. SEM uses electrons for imaging, much as a light microscope uses visible light.

- Transmission electron microscopy (TEM) is a high magnification measurement technique that images the transmission of a beam of electrons through a sample.

- Energy-dispersive X-ray spectroscopy (EDS, EDX, EDXS or XEDS), sometimes called energy dispersive X-ray analysis (EDXA) or energy dispersive X-ray microanalysis (EDXMA), is an analytical technique used for the elemental analysis or chemical characterization of a sample.

## EXERCISE

1. Draw and explain FTIR.
2. Draw and explain XRD.
3. Draw and explain AFM.
4. Draw and explain SEM.
5. Draw and explain TEM.
6. Explain EDAX Applications and interpretation of results.
7. What are the different Emerging nano material and devices?
8. Draw and explain nano tubes.
9. Draw and explain Nano rods.
10. Describe LB technique.
11. Explain Soft lithography Microwave assisted synthesis.
12. Explain Self-assembly.

◈ ◈ ◈

## MODEL QUESTION PAPERS FOR
## End-Semester Examination

**Paper - I**

**Time : 3 Hours**                                                                 **Max. Marks : 60**

**Instructions to the candidates :**

(1)  Each Question carries 12 Marks.

(2)  Attempt any five questions to the following.

(3)  Illustrate your answers with neat sketches, diagram etc., wherever necessary.

(4)  If some part or parameter is noticed to be missing, you may appropriately assume it and should mention it clearly.

| | | |
|---|---|---|
| **1.** **(a)** Discuss v/c-v characterization. | | **(6)** |
| **(b)** (i) Explain water fabrication techniques. | | **(3)** |
| (ii) Recognize the basic steps of CMOS. | | **(3)** |
| **2.** **(a)** Draw and explain Bond theory of Solids. | | **(6)** |
| **(b)** Describe Brillouin zones. | | **(6)** |
| **3.** **(a)** Draw and explain FinFets. | | **(6)** |
| **(b)** (i) Explain system integration limits. | | **(3)** |
| (ii) Explain shrink down approaches. | | **(3)** |
| **4.** **(a)** Explain Quantum blockade. | | **(6)** |
| **(b)** (i) State Applications of transport devices. | | **(3)** |
| (ii) Write a note on Graphene. | | **(3)** |
| **5.** **(a)** Draw and explain Vertical transistors. | | **(6)** |
| **(b)** Explain Germanium and compound semiconductors. | | **(6)** |
| **6.** **(a)** Draw and explain SEM. | | **(6)** |
| **(b)** (i) Draw and explain Nano rods. | | **(3)** |
| (ii) Draw and explain XRD. | | **(3)** |

## Paper - II

**Time : 3 Hours**        **Max. Marks : 60**

**Instructions to the candidates :**

(1) Each Question carries 12 Marks.

(2) Attempt any five questions to the following.

(3) Illustrate your answers with neat sketches, diagram etc., wherever necessary.

(4) If some part or parameter is noticed to be missing, you may appropriately assume it and should mention it clearly.

---

**1.**   **(a)** Write a short note on the CMOS technology.      **(6)**

     **(b)** (i)   Explain the crystal growth and wafer fabrication.      **(3)**

          (ii)   State and explain the construction of mass field effect transistor.      **(3)**

**2.**   **(a)** Describe Schrodinger equation.      **(6)**

     **(b)** Explain Kronig-Penny Model.      **(6)**

**3.**   **(a)** Describe CMOS Scaling.      **(6)**

     **(b)** Draw and explain operation of Vertical MOSFETs.      **(6)**

**4.**   **(a)** Describe Coulomb dots.      **(6)**

     **(b)** (i)   Explain Carbon nanotube electronics.      **(3)**

          (ii)   What is Atomistic simulation? Explain.      **(3)**

**5.**   **(a)** Draw and explain Vertical transistors.      **(6)**

     **(b)** What are the different Properties of schottky functions on Silicon ?      **(6)**

**6.**   **(a)** Draw and explain FTIR.      **(6)**

     **(b)** Explain EDAX Applications and interpretation of results.      **(6)**

---